Lecture Notes in Electrical Engineering

Volume 109

For further volumes:
http://www.springer.com/series/7818

Arnaldo D'Amico • Corrado Di Natale
Lucia Mosiello • Giovanna Zappa
Editors

Sensors and Microsystems

AISEM 2011 Proceedings

Springer

Editors
Arnaldo D'Amico
University of Rome Tor Vergata
Rome, Italy
damico@eln.uniroma2.it

Corrado Di Natale
University of Rome Tor Vergata
Rome, Italy
dinatale@uniroma2.it

Lucia Mosiello
ENEA
Italian National Agency
for New Technologies
Energy and the Environment
Rome, Italy
mosiello@casaccia.enea.it

Giovanna Zappa
ENEA
Italian National Agency
for New Technologies
Energy and the Environment
Rome, Italy
Giovanna.zappa@enea.it

ISSN 1876-1100
e-ISSN 1876-1119
ISBN 978-1-4899-9944-3
ISBN 978-1-4614-0935-9 (eBook)
DOI 10.1007/978-1-4614-0935-9
Springer New York Dordrecht Heidelberg London

Softcover reprint of the hardcover 1st edition 2012

Printed on acid-free paper

Springer is part of Springer Science+Business Media (www.springer.com)

Foreword

The Italian Association of Sensors and Microsystems was funded in 1995 with the scope of promoting the research and diffusing the culture of sensors in Italy. The major outcome of the association is the national conference that runs yearly from the first edition in 1996.

The 16th conference was organized by the "Italian National Agency for New Technologies, Energy and Sustainable Economic Development" (ENEA) and was held from 7 to 9 February, 2011 at the ENEA Casaccia Research Centre.

This edition gave special importance to the Metrology and Quality of Measurement. The opening presentation on *Metrology and Measurement Reliability* was held by Prof. R.F. Laitano and the third day of the conference started with a presentation on *Quality of Chemical and Biological Measurement.*

The conference was also the occasion to pay a tribute to the memory of late Prof. Giuliano Martinelli that played a significant role in the community of sensors and microsystems. Memorial lectures were given by the following distinguished colleagues: Joan Morante (University of Barcelona), Udo Weimar and Nicolae Barsan (University of Tübingen) and Giorgio Sberveglieri (University of Brescia). The session was complemented by talks given by the collaborators of Prof. Martinelli at the University of Ferrara: Maria Cristina Carotta, Vincenzo Guidi, and Cesare Malagù. Our heartfelt thanks come to the speakers that honoured, with their presence, the scientific life of Giuliano Martinelli.

The conference numbered about 230 authors from 76 different affiliations with a remarkable participation of the Academic Community, several institutes of the National Research Council (CNR), many research groups of ENEA and a significant presence of sensors companies. In an interdisciplinary approach many aspects of the disciplines have been covered, ranging from materials science, chemistry, applied physics, electronic engineering and biotechnologies.

Special thanks are given to Eng. Giovanni Lelli, Commissioner of ENEA, for his involvement and encouragement and to Eng. Marco Citterio, Director of Casaccia Research Centre, and Secretary Staff for the commitment to the conference organisation. We would like also to thank all the authors who contributed with their papers to the success of the conference.

This Book gathers a selection of the papers presented at the conference; it contains contributions from both academic and industrial researchers providing a unique perspective on the research and development of sensors, microsystems and related technologies in Italy. The scientific value of the papers also offers an invaluable source to analysts intending to survey the contribution of Italian researchers in the field of sensors and microsystems.

Rome, Italy

Arnaldo D'Amico
Corrado Di Natale
Lucia Mosiello
Giovanna Zappa

Contents

Contributors

A. Adami Bio-MEMS, FBK, Trento, Italy

D. Aloisio Dipartimento di Fisica della Materia e Ingegneria Elettronica, DFMIE, Università degli Studi di Messina, Messina, Italy

M. Alvisi ENEA, Brindisi Technical Unit for Technologies of Materials, Brindisi, Italy

M. Amodio Department of Chemistry, University of Bari, Bari, Italy

P. Antonucci Università "Mediterranea" di Reggio Calabria – Facoltà di Ingegneria – Reggio Calabria, Reggio Calabria, Italy

P. Aquaro Vibo Valentia, Italy

A. Arena Dipartimento di Fisica della Materia e Ingegneria Elettronica, DFMIE, Università degli Studi di Messina, Messina, Italy

M.N. Armenise Dipartimento di Elettrotecnica ed Elettronica, Politecnico di Bari, Bari, Italy

P. Aversa ENEA, Brindisi Technical Unit for Technologies of Materials, Brindisi, Italy

F. Baldini Institute of Applied Physics "Nello Carrara", National Research Council of Italy, Sesto Fiorentino (FI), Italy

L. Barboni Department of Biophysical and Electronic Engineering, University of Genova, Genoa, Italy

G. Barillaro Dipartimento di Ingegneria dell'informazione: Elettronica, Informatica, Telecomunicazioni, Università di Pisa, Pisa, Italy

G. Barlocchi STMicroelectronics, Cornaredo, Milan, Italy

F. Barreca Advanced and Nano Materials Research s.r.l., Messina, Italy

A. Baschirotto Department of Physics, University of Milano Bicocca, Milano, Italy

M. Bassu Dipartimento di Ingegneria dell'informazione: Elettronica, Informatica, Telecomunicazioni, Università di Pisa, Pisa, Italy

R. Bernini Istituto per il Rilevamento Elettromagnetico dell'Ambiente – Consiglio Nazionale delle Ricerche, Napoli, Italy

A. Bonavita Department of Industrial Chemistry and Materials Engineering, University of Messina, Messina, Italy

M. Brattoli Department of Chemistry, University of Bari, Bari, Italy

M. Brenci Institute of Applied Physics "Nello Carrara", National Research Council of Italy, Sesto Fiorentino (FI), Italy

F. Brunetti Department of Electronic Engineering, Engineering University of Rome Tor Vergata, Rome, Italy

P. Bruschi Dipartimento di Ingegneria dell'informazione: Elettronica, Informatica, Telecomunicazioni, Università di Pisa, Pisa, Italy

F. Butti Dipartimento di Ingegneria dell'informazione: Elettronica, Informatica, Telecomunicazioni, Università di Pisa, Pisa, Italy

L. Campanella Department of Chemistry, University of Rome "La Sapienza", Rome, Italy

C.M. Campanella Dipartimento di Elettrotecnica ed Elettronica, Politecnico di Bari, Bari, Italy

G. Campobello Dipartimento di Fisica della Materia e Ingegneria Elettronica, DFMIE, Università degli Studi di Messina, Messina, Italy

G. Cannata Department of Communication Computer and System Sciences, University of Genoa, Genoa, Italy

G. Cannatà Dipartimento di Fisica della Materia e Ingegneria Elettronica, DFMIE, Università degli Studi di Messina, Messina, Italy

C. Cannilla CNR-TAE "Nicola Giordano", Messina, Italy

C. Cantalini Department of Chemistry, Chemical Engineering and Materials, University of L'Aquila, L'Aquila, Italy

R. Capuano Department of Electronic Engineering, University of Rome Tor Vergata, Rome, Italy

M. Capurro Research Center on Materials Science and Technology, University of Genoa, Genoa, Italy
Department of Civil, Environmental and Architectural Engineering, University of Genoa, Genoa, Italy

P. Cardiano Analytical Chemistry and Physical Chemistry, University of Messina, Messina, Italy

L. Carlucci Department of SCTSBA and Department of Chemistry and Pharmacy Technology, University of Rome “La Sapienza”, Rome, Italy

M. Caruso Department of Chemical Sciences and Technologies, University of Rome Tor Vergata, Rome, Italy

V. Casola Computer Science and Systems Department, University of Napoli Federico II, Naples, Italy

G. Cassano ENEA, Brindisi Technical Unit for Technologies of Materials, Brindisi, Italy

G. Castillo Department of Nuclear Physics and Biophysics, Comenius University, Bratislava, Slovakia

A. Catini Department of Electronic Engineering, University of Rome “Tor Vergata”, Rome, Italy

D. Cattin Department of Management and Engineering, University of Padova, Vicenza, Italy

A. Cervi Department of Physics, University of Ferrara, Ferrara, Italy

F. Chiavaioli Department of Information Engineering, University of Siena, Siena, Italy

L. Ciaccheri CNR IFAC, Sesto Fiorentino (FI), Italy

C. Ciminelli Dipartimento di Elettrotecnica ed Elettronica, Politecnico di Bari, Bari, Italy

C. Ciofi Dipartimento di Fisica della Materia e Ingegneria Elettronica, DFMIE, Università degli Studi di Messina, Messina, Italy

P. Cirillo Research Center on Materials Science and Technology, University of Genoa, Genoa, Italy
Department of Civil, Environmental and Architectural Engineering, University of Genoa, Genoa, Italy

C. Collini Bio-MEMS, FBK, Trento, Italy

D. Compagnone Dipartimento di Scienze degli Alimenti, Universita´ degli studi di Teramo, Mosciano Sant’Angelo (TE), Italy

I. Concina CNR-IDASC SENSOR Laboratory and Brescia University, Brescia, Italy

F. Conso Department of Electrical Engineering, University of Pavia, Pavia, Italy

A. Costanzo Department of Electronics, Computer Science and Systems, University of Bologna, Cesena, Italy

M. Crisma ICB, Padova Unit, CNR, Department of Chemistry, University of Padova, Padova, Italy

G. Currò Advanced and Nano Materials Research s.r.l., Messina, Italy

A. D'Amico Department of Electronic Engineering, University of Rome "Tor Vergata", Rome, Italy

M. Dachà Center for Integrated Research - CIR, Unit of Food and Nutrition, "Università Campus Bio-Medico di Roma", via Álvaro del Portillo 21, 00128 Rome, Italy

R.S. Dahiya Bio-MEMS, FBK, Trento, Italy

G. De Gennaro Department of Chemistry, University of Bari, Bari, Italy

A. De Marcellis Department of Electrical and Information Engineering, University of L'Aquila, L'Aquila, Italy

S. De Vito Basic Materials and Devices Department, ENEA – National Agency for New Technologies, Energy and Sustainable Development, Portici (NA), Italy

C. Di Carlo Department of Electrical and Information Engineering, University of L'Aquila, L'Aquila, Italy

G. Di Francia ENEA Centro Ricerche Portici, Portici (NA), Italy

C. Di Natale Department of Electronic Engineering, University of Rome Tor Vergata, Rome, Italy

C. Di Palma Computer Science and Systems Department, University of Napoli Federico II, Naples, Italy

F. Dini Department of Electronic Engineering, University of Rome Tor Vergata, Rome, Italy

N. Donato Dipartimento di Fisica della Materia e Ingegneria Elettronica, DFMIE, Università degli Studi di Messina, Messina, Italy

A. Donato Università "Mediterranea" di Reggio Calabria – Facoltà di Ingegneria – Reggio Calabria, Reggio Calabria, Italy

L. Dugo Center for Integrated Research - CIR, Unit of Food and Nutrition, "Università Campus Bio-Medico di Roma", via Álvaro del Portillo 21, 00128 Rome, Italy

M.L. Ermini Dipartimento di Chimica "Ugo Schiff", Università degli Studi di Firenze, Sesto F.no (FI), Italy

M. Falasconi CNR-IDASC SENSOR Laboratory, Brescia, Italy

G. Fattoruso Basic Materials and Devices Department, ENEA – National Agency for New Technologies, Energy and Sustainable Development, Portici (NA), Italy

P. Fava Department of Agricultural and Food Sciences, Modena and Reggio Emilia University, Reggio Emilia, Italy

G. Favero Department of SCTSBA and Department of Chemistry and Pharmacy Technology, University of Rome "La Sapienza", Rome, Italy

G. Ferri Department of Electrical and Information Engineering, University of L'Aquila, L'Aquila, Italy

F. Formaggio ICB, Padova Unit, CNR, Department of Chemistry, University of Padova, Padova, Italy

G. Fusella Dipartimento di Scienze degli Alimenti, Universita´ degli studi di Teramo, Mosciano Sant'Angelo (TE), Italy

M. Galeano Dipartimento di Fisica della Materia e Ingegneria Elettronica, DFMIE, Università degli Studi di Messina, Messina, Italy

E. Gatto Department of Chemical Sciences and Technologies, University of Rome Tor Vergata, Rome, Italy

L. Giancaterini Department of Chemistry, Chemical Engineering and Materials, University of L'Aquila, L'Aquila, Italy

L. Giannelli Hospitex Diagnostics srl, Sesto Fiorentino (FI), Italy

A. Gianni Department of Electronic Engineering, University of Rome Tor Vergata, Rome, Italy

A. Giberti Department of Physics, University of Ferrara, Ferrara, Italy
MIST E-R S.C.R.L, Bologna, Italy

C. Granata Istituto di Cibernetica "E. Caianiello" del Consiglio Nazionale delle Ricerche, Pozzuoli (Napoli), Italy

M. Grassi Department of Electrical Engineering, University of Pavia, Pavia, Italy

E. Hamidi-Asl Electroanalytical Chemistry Research Laboratory, Department of Analytical Chemistry, University of Mazandran, Babolsar, Iran

T. Hianik Department of Nuclear Physics and Biophysics, Comenius University, Bratislava, Slovakia

H. Ilkhani Department of Chemistry, University of Guilan, Rasht, Iran
Center of Excellence in Electrochemistry, University of Tehran, Tehran, Iran

C.D. Kohl Institute of Applied Physics, Justus-Liebig-University, Giessen, Germany

I. Lamberti Department of Biology, University of Roma Tre, Roma, Italy

S. Laschi Dipartimento di Chimica, Università degli Studi di Firenze, Sesto Fiorentino, Italy

M. Latino Department of Chemical Sciences and Technologies, University of Rome Tor Vergata, Rome, Italy

S. Lo Schiavio Analytical Chemistry and Physical Chemistry, University of Messina, Messina, Italy

L. Lorenzelli Bio-MEMS, FBK, Trento, Italy

C. Malagù Department of Physics, University of Ferrara, Ferrara, Italy

P. Malcovati Department of Electrical Engineering, University of Pavia, Pavia, Italy

C. Marichy Department of Chemistry, CICECO, University of Aveiro, Aveiro, Portugal

G. Marrazza Dipartimento di Chimica, Università degli Studi di Firenze, Sesto Fiorentino, Italy

E. Martinelli Department of Electronic Engineering, University of Rome Tor Vergata, Rome, Italy

E. Martini Department of Chemistry, University of Rome "La Sapienza", Rome, Italy

M. Mascini Dipartimento di Chimica, Università degli Studi di Firenze, Sesto Fiorentino, Italy

D. Masotti Department of Electronics, Computer Science and Systems, University of Bologna, Bologna, Italy

E. Massera ENEA Centro Ricerche Portici, Portici (NA), Italy

U. Mastromatteo STMicroelectronics, Cornaredo, Milan, Italy

F. Mazzei Department of SCTSBA and Department of Chemistry and Pharmacy Technology, University of Rome "La Sapienza", Rome, Italy

A.A. Mencaglia CNR IFAC, Sesto Fiorentino (FI), Italy

G. Metta RBCS, Italian Institute of Technology, Genova, Italy

M.L. Miglietta ENEA Centro Ricerche Portici, Portici (NA), Italy

A.G. Mignani CNR IFAC, Sesto Fiorentino (FI), Italy

A. Minardo Dipartimento di Ingegneria per l'Informazione, Seconda Università di Napoli, Aversa, Italy

M. Minunni Dipartimento di Chimica "Ugo Schiff", Università degli Studi di Firenze, Sesto Fiorentino, Firenze, Italy

V. Modafferi Università "Mediterranea" di Reggio Calabria – Facoltà di Ingegneria – Reggio Calabria, Reggio Calabria, Italy

L. Mosiello ENEA, Italian National Agency for New Technologies, Energy and the Environment, Rome, Italy

M. Mugnaini Department of Information Engineering, University of Siena, Siena, Italy

A. Nannini Dipartimento di Ingegneria dell'informazione, Università di Pisa, Pisa, Italy

I. Nasti ENEA Centro Ricerche Portici, Portici (NA), Italy

G. Neri Department of Industrial Chemistry and Materials Engineering, University of Messina, Messina, Italy

R. Oboe Department of Management and Engineering, University of Padova, Vicenza, Italy

I. Palchetti Dipartimento di Chimica, Università degli Studi di Firenze, Sesto Fiorentino, Italy

G. Panzera Università "Mediterranea" di Reggio Calabria – Facoltà di Ingegneria – Reggio Calabria, Reggio Calabria, Italy

R. Paolesse Department of Chemical Sciences and Technologies, University of Rome Tor Vergata, Rome, Italy

M. Pardo Institute of Applied Mathematics and Information Technology, CNR, Genova, Italy

G. Pennazza Center for Integrated Research - CIR, Unit of Electronics for sensor systems, "Università Campus Bio-Medico di Roma", via Álvaro del Portillo 21, 00128 Rome, Italy

M. Penza ENEA, Brindisi Technical Unit for Technologies of Materials, Brindisi, Italy

F. Pieri Dipartimento di Ingegneria dell'informazione, Università di Pisa, Pisa, Italy

N. Pinna Department of Chemistry, CICECO, University of Aveiro, Aveiro, Portugal

M. Piotto CNR IEIIT, Pisa, Italy

E. Piscopiello ENEA, Brindisi Technical Unit for Technologies of Materials, Brindisi, Italy

T. Polichetti ENEA Centro Ricerche Portici, Portici (NA), Italy

A. Porchetta Department of Chemical Sciences and Technologies, University of Rome Tor Vergata, Rome, Italy

A. Pulvirenti Department of Agricultural and Food Sciences, Modena and Reggio Emilia University, Reggio Emilia, Italy

F. Ricciardella ENEA Centro Ricerche Portici, Portici (NA), Italy

A. Romani Department of Electronics, Computer Science and Systems, University of Bologna, Cesena, Italy

S. Romano ENEA Centro Ricerche Portici, Portici (NA), Italy

R. Rossi ENEA, Brindisi Technical Unit for Technologies of Materials, Brindisi, Italy

B. Ruggiero Istituto di Cibernetica "E. Caianiello" del Consiglio Nazionale delle Ricerche, Pozzuoli (Napoli), Italy

V. Russino Dipartimento di Ingegneria dell'informazione, Università di Pisa, Pisa, Italy

M. Russo Istituto di Cibernetica "E. Caianiello" del Consiglio Nazionale delle Ricerche, Pozzuoli (Napoli), Italy

G. Saitta Dipartimento di Fisica della Materia e Ingegneria Elettronica, DFMIE, Università degli Studi di Messina, Messina, Italy

E. Sangiorgi Department of Electronics, Computer Science and Systems, University of Bologna, Cesena, Italy

M. Santonico Department of Electronic Engineering, University of Rome "Tor Vergata", Rome, Italy

C.I. Saulle Dipartimento di Scienze degli Alimenti, Universita´ degli studi di Teramo, Mosciano Sant'Angelo (TE), Italy

V. Sberveglieri Department of Agricultural and Food Sciences, Modena and Reggio Emilia University, Reggio Emilia, Italy

G. Scandurra Dipartimento di Fisica della Materia e Ingegneria Elettronica, DFMIE, Università degli Studi di Messina, Messina, Italy

S. Scarano Dipartimento di Chimica "Ugo Schiff", Università degli Studi di Firenze, Sesto F.no (FI), Italy

M. Schioppa ENEA, Brindisi Technical Unit for Technologies of Materials, Brindisi, Italy

A. Secchi SELEX Sistemi Integrati S.p.A, Rome, Italy

L. Seminara Department of Biophysical and Electronic Engineering, University of Genoa, Genoa, Italy

E. Serra ENEA, Brindisi Technical Unit for Technologies of Materials, Brindisi, Italy

S. Serrano Dipartimento di Fisica della Materia e Ingegneria Elettronica, DFMIE, Università degli Studi di Messina, Messina, Italy

S. Spadaro Advanced and Nano Materials Research s.r.l.,Messina, Italy

D. Spadaro Department of Industrial Chemistry and Materials Engineering, University of Messina, Messina, Italy

L.M. Strambini Dipartimento di Ingegneria dell'informazione: Elettronica, Informatica, Telecomunicazioni, Università di Pisa, Pisa, Italy

S. Surdo Dipartimento di Ingegneria dell'informazione: Elettronica, Informatica, Telecomunicazioni, Università di Pisa, Pisa, Italy

D. Suriano ENEA, Brindisi Technical Unit for Technologies of Materials, Brindisi, Italy

M.A. Tagliente ENEA, Brindisi Technical Unit for Technologies of Materials, Brindisi, Italy

M. Tartagni Department of Electronics, Computer Science and Systems, University of Bologna, Cesena, Italy

M. Tiemann Faculty of Science, Department of Chemistry, University of Paderborn, Paderborn, Germany

M. Tomassetti Department of Chemistry, University of Rome "La Sapienza", Rome, Italy

S. Tombelli Dipartimento di Chimica "Ugo Schiff", Università degli Studi di Firenze, Sesto F.no (FI), Italy

C. Toniolo ICB, Padova Unit, CNR, Department of Chemistry, University of Padova, Padova, Italy

L. Tortora Department of Chemical Sciences and Technologies, University of Rome Tor Vergata, Rome, Italy

F. Tortorella Information Engineering Department, Università di Cassino, Cassino (FR), Italy

L. Trizio Department of Chemistry, University of Bari, Bari, Italy

C. Trono Institute of Applied Physics "Nello Carrara", National Research Council of Italy, Sesto Fiorentino (FI), Italy

M. Valle Department of Biophysical and Electronic Engineering, University of Genova, Genoa, Italy
Research Center on Materials Science and Technology, University of Genova, Genoa, Italy

M. Venanzi Department of Chemical Sciences and Technologies, University of Rome Tor Vergata, Rome, Italy

A. Vettoliere Istituto di Cibernetica "E. Caianiello" del Consiglio Nazionale delle Ricerche, Pozzuoli (Napoli), Italy

F.F. Villa STMicroelectronics, Cornaredo, Milan, Italy

T. Wagner Faculty of Science, Department of Chemistry, University of Paderborn, Paderborn, Germany

T. Waitz Institute of Inorganic Chemistry, Chemical Didactics, Georg-August-Universität, Göttingen, Germany

M. WIllinger Department of Chemistry, CICECO, University of Aveiro, Aveiro, Portugal

L. Zeni Dipartimento di Ingegneria per l'Informazione, Seconda Università di Napoli, Aversa, Italy

Part I
Biosensors

Chapter 1
Determination of Immunoglobulins G in Human Serum and Cow Milk Using a Direct Immunological Method Based on Surface Plasmon Resonance

Mauro Tomassetti, Elisabetta Martini, Luigi Campanella, Luciano Carlucci, Gabriele Favero, and Franco Mazzei

A new method for IgG analysis in real matrixes, such as serum and several types of fresh or powdered milks was studied using a surface plasmon resonance (SPR) apparatus in the Kretschmann configuration, obtaining satisfactory results.

1 Introduction

Within the framework of research carried out by our team aimed at developing new immunological methods to determine proteins such as Immunoglobulins G in different biological matrixes, such as serum and milk, tests performed in previous researches were based on several different immunosensors and using different transducer types: potentiometric (ISE for NH_3) [1], amperometric (amper. – tyrosinase enzyme sensor) [2], or screen printed electrodes for hydrogen peroxide [3]; our team is currently testing the feasibility of constructing a new immunosensor for IgG determination based on surface plasmon resonance (SPR). Different construction techniques and measurement geometries were used in previous researches, involving also different enzymatic markers. Furthermore, "competitive" immunological procedures were used in most cases. Conversely, the SPR (surface plasmon resonance) transduction technique used in the present research allowed "direct" measurement.

M. Tomassetti (✉) • E. Martini • L. Campanella
Department of Chemistry, University of Rome "La Sapienza", Rome, Italy
e-mail: mauro.tomassetti@uniroma1.it

L. Carlucci • G. Favero • F. Mazzei
Department of SCTSBA and Department of Chemistry and Pharmacy Technology, University of Rome "La Sapienza", Rome, Italy

A. D'Amico et al. (eds.), *Sensors and Microsystems: AISEM 2011 Proceedings*, Lecture Notes in Electrical Engineering 109, DOI 10.1007/978-1-4614-0935-9_1,

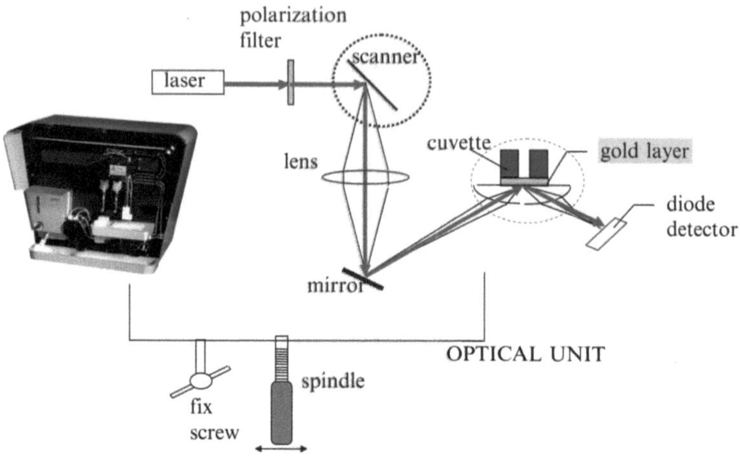

Fig. 1.1 Surface plasmon resonance instrumentation

2 Equipment

The Surface Plasmon Resonance (SPR) experiments for IgG determination were performed using an ESPRIT instrument (Echo Chemie B.V., Utrecht, The Netherlands, shown in Fig. 1.1. In this device, based on the Kretschmann configuration, the intensity of the reflected light is minimum at the angle of resonance. Angles can be measured over a range of 4° using a diode detector. The angle of incidence is varied using an oscillating mirror, which rotates over an angle of 5°, directing a polarized laser (wavelength 670 nm) on a surface (1 ×2 mm) of the disk which is the sensor through a glass semi-cylindrical prism. During each cycle the reflectivity of the mirror is measured for each movement, with a resolution for this configuration of 1 m°.

3 Method

In the experiments a sensor (Xantec Bioanalytical), consisting of a glass disk covered with a 50 nm thick Au layer superimposed on a 1.5 nm Ti layer required for the purpose of adhesion was mounted in a Teflon SPR cell. Before use, the Au surface was cleaned with a solution of concentrated H_2SO_4 and 33% H_2O_2 in a 3:1 ratio and the resulting oxide layer removed by immersion in absolute ethanol for 10 min. The Au surface, which was cleaned in this way before use, was modified by dipping it into a millimolar alcohol solution of mercaptoundecanoic acid, thus obtaining a SAM (self assembled monolayer) that makes it possible to chemically bond the selected antibody (anti-IgG) to the surface by means of a reaction with carbodiimide and succinimide. When the disk thus prepared was placed

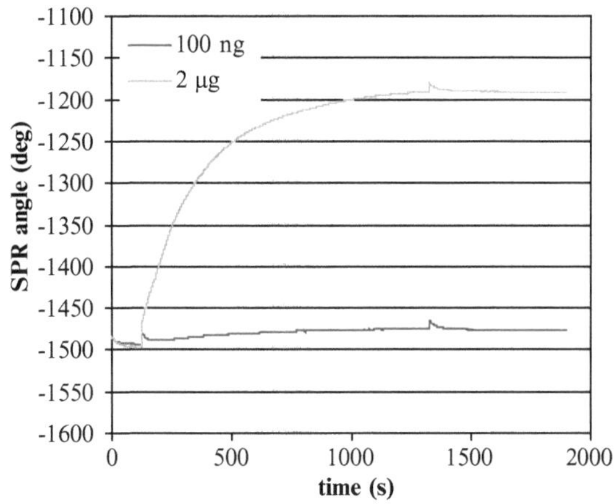

Fig. 1.2 A typical curve obtained with surface plasmon resonance for a fixed concentration of immunoglobulin G of the order of 2 μg L^{-1}

in contact with a solution containing antigen to determine IgG on its surface to form the antibody, thereby changing the resonance angle, which will be a function of the concentration of IgG in the solution. This produced a series of curves for different concentrations of IgG; an example of a typical curve obtained is shown in Fig. 1.2.

4 Results and Discussion

A calibration curve was thus constructed (see Fig. 1.3) in which also the equation of the straight line obtained is reported. The method display a linear range of between 3 and 30 nmol L^{-1} of IgG and an LOD of 1.0 nmol L^{-1}. The method measurement time, which entails the use of surface plasmon resonance, is about half an hour or lower. Also the value of the affinity constant was estimated: the K_{aff} value was found to be of the order of 10^7 L mol^{-1}.

Finally, the method (SPR) was applied to the determination of IgG concentration in human serum and cow's milk, which were respectively found to be 3,820 and 1,070 mg L^{-1}, with an RSD% <10. Lastly, this method was used for the determination of IgG concentration in goat or buffalo milk and in samples of powdered milk for babies. The values obtained are shown in Table 1.1.

It should be noted that the values are very different for the different animal species, regardless of whether fresh whole milk, frozen milk, or powdered milk are analyzed. A significantly low concentration was evidenced, for instance, when stored frozen buffalo milk was analyzed. Of course, the latter showed a much lower concentration than the sample of fresh cow milk.

Fig. 1.3 Calibration curve obtained by the direct method based on surface plasmon resonance (SPR) for IgG determination

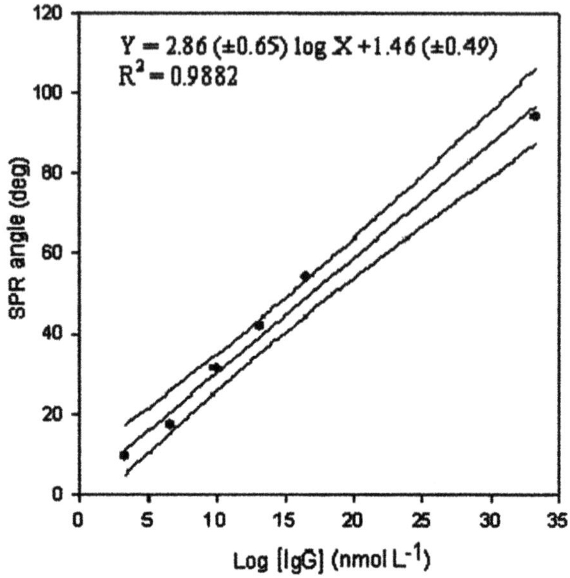

Table 1.1 Determination of the IgG concentration values in milk and in human serum. Values expressed in mg L^{-1}

Real samples	Surface plasmon resonance IgG concentration found (mg L^{-1}); n = 5; RSD% ≤ 10
Cow milk	1,070 (high quality fresh milk)
Goat milk	1,285 (whole)
Buffalo milk	450 (stored in freezer)
Powdered milk samples	650–1,000
Serum	3,820

5 Conclusions

This new "direct" method (SPR) reduces the time needed to perform the analysis, even if the same LOD value is obtained as for the classical immunosensor methods using the competition procedure [4]. It also allows the suitable determination of value of IgG concentration in several real matrices, such as human serum and goat, cow, or powdered milk samples.

Acknowledgment This work was funded by Sapienza University of Rome, "Ateneo Project".

References

1. Campanella L, Martini E, Tomassetti T (2008) Determination of HIgG and anti-HIgG using a single potentiometric immunosensor and two different "competitive methods": application to the analysis of globulin G in human serum. Sensor Actuator B 130:520–530
2. Campanella L, Lelo D, Martini E, Tomassetti M (2008) Immunoglobulin G determination in human serum and milk using an immunosensor of new conception fitted with an enzyme probe as transducer. Sensors 8:6727–6746
3. Campanella L, Favero G, Martini E, Tomassetti M (2009) Sensors and Microsystems. In: D'Amico A, Di Natale C, Martinelli E, Paolesse R (eds) Proceedings of the 13th Italian conference, Rome, 19–21 February 2008. World Scientific Printers, Singapore, pp 46–50
4. Campanella L, Martini E, Tomassetti M (2008) New immunosensor for lactoferrin determination in human milk and several pharmaceutical dairy milk products recommended for the unweaned diet. J Pharm Biomed Anal 48:278–287

Chapter 2
Erythropoietin Detection: A Biosensor Approach*

S. Scarano, M.L. Ermini, S. Tombelli, M. Mascini, and M. Minunni

Erythropoietin is a glycoproteic hormone of 165 aminoacids belonging to hemopoietic growth factors. Recombinant EPO has been introduced in early '80 to treat patients with severe anemia and to reduce side effects other EPO produced in eukaryotes have been introduced and different analogs produced over years. Since 2004, when EPO alfa and beta patents were over other molecules have appeared as EPO derivatives. EPO results in the prohibited list of World Anti-doping organization (WADA) and their abuse should be controlled in sport. EPO analysis is quite difficult since the molecule has relatively short half-life, numerous isoforms and many analogs are present on the market. We will report about EPO detection in urine samples. To achieve this, antibodies were used as recognition elements in sensing developments using Surface Plasmon Resonance (SPR) as transduction principle.

1 Introduction

1.1 The Analytical Problem

Erythropoietin (EPO) is a glycoproteic hormone of 165 aminoacids (MW circa 30 KDa), belonging to hemopoietic growth factors. The total molecular mass depends on the glycosylation degree of the protein, which is quite variable, and originates from posttranslational modification, and represents the only difference between the isoforms of human EPO, all biologically active, differing in the

---

*This work is supported by World Anti-Doping Agency

S. Scarano • M.L. Ermini • S. Tombelli • M. Mascini • M. Minunni (✉)
Dipartimento di Chimica "Ugo Schiff", Università degli Studi di Firenze, Sesto F.no (FI), Italy
e-mail: maria.minunni@unifi.it

A. D'Amico et al. (eds.), *Sensors and Microsystems: AISEM 2011 Proceedings*,
Lecture Notes in Electrical Engineering 109, DOI 10.1007/978-1-4614-0935-9_2,

isolectric point (IP) in the range 3,8 and 4,7 [1, 2]. EPO is found in human serum and urine. Protein concentration are 2.5 pM in serum and in 0.4 pM urine respectively. Recombinant EPO has been introduced in early '80 to treat patients with severe anemia and to reduce side effects other EPO produced in eukaryotes have been introduced and different analogs produced over years and their abuse should be controlled in sport. Since 2000 at Sydney Olympic games, EPO has been analyzed first in serum and then in urine for confirmation analysis, and since then direct analysis of EPO in urine should be mandatory since 2003, following WADA indications. Current analytical methods are based on electrofocusing, coupled to double blotting [3]. EPO analysis is quite difficult since the molecule has relatively short half-life, numerous isoforms and many analogs are present on the market.

Preliminary results of affinity-based sensing for EPO detection in urine samples will be here reported. To achieve this, antibodies were used as recognition elements in sensing development using Surface Plasmon Resonance (SPR) as transduction principle.

2 Experimental

2.1 *Immunosensor Development*

Reagents used for immobilization of anti-EPO antibodies are: N-idrossisuccinimide (NHS) from Fluka (Milan, Italy), N-(3-dimetilamminopropil)-N-ethylcarbodiimide (EDAC) from Calbiochem (associated to Merck Darmstadt, Germany), ethanolamine (EA) from Sigma Aldrich (Milano, Italy). Monoclonal Antibodies (mAb) specific for recombinant EPO (rEPO) and urinary EPO (uEPO) were from R&D System (Minneapolis, USA): mAb 287 (clone 9C21D11) specific recombinant EPO (rEPO), produced in mice, lyophilized was diluted and aliquoted in PBS; mAb 2871 (clone AE7A5) non discriminating among rEPO and uEPO, produced in mice was purchased in PBS solution. EPO analytes, both human recombinant (rhEPO) and endogenous human urinary (uhEPO) were from NIBSC (National Institute for Biological Standard and Control, UK). For sensor regeneration HCl, NaOH and glycine (gly) solutions at different concentrations were used. All reagents for the buffers (binding buffer PBS pH 7.4) were from Merck (Darmstadt, Germany). For testing sensor specificity human serum albumin (HSA, 97–99% purity), dithiothreitol (DTT) were from Sigma (Milan, Italy). Instrumentation: BIAcore X® (GE Healthcare Europe).

2.2 *CM5 Chip™ Modification*

The flow cells on the CM5 (GE Healthcare Europe, Milan, Italy) were modified with mAb 287 (clone 9C21D11) on flow cell 1 (fc1) and mAb 2871 (clone AE7A5) flow cell 2 (fc2), following indication by Bulukin et al [4].

Immobilization parameters: flow 5 μL/min and temperature T = 25°C. Activation solution: NHS 50 mM and EDAC 200 mM in distilled water (contact time: 7 min (35 μL), followed by 100 ppm antibody injection in acetate buffer (75 μL, contact time 15 min). 35 μL di ethanolamine (EA) 1 M pH = 8.0 is finally injected to saturate activated sites.

3 Results

3.1 *Optimization of the Surface Regeneration*

To allow measurement reproducibility, after each measurement cycle the immunocomplex should be dissociated and the signal has to come back to the initial base line. In other words, to study the regeneration efficiency the signal after the antibody-antigen interaction and the subsequent complex dissociation is recorded. Ideally a constant signal (base line) after each regeneration, should be obtained. Here three different solutions were applied for dissociating the immunocomplex (after injection of 47 ppb rhEPO): HCl, NaOH and Gly pH 1.9, for different contact time with the surface (15, 25 and 40 min). Glycine solution was selected as regenerating agent, since HCl and NaOH were not effective. Thus 5 mM glycine pH 1.9 was chosen for regeneration with 40 s contact time, allowing 50 measurements on the same chip.

3.2 *Sensor Calibration with rh-EPO Buffer Solutions*

For the calibration, standard solutions in the rh-Epo analyte range 94; 47, 23, 15, and 10 ppb, were prepared (Fig.2.1). A good sensitivity (DL = 15 ng/mL) was found, with a reproducibility, expressed as coefficient of variation CV% of 18%. However low selectivity was found, since response was observed also with uEPO, thus we calculated the ratio between the sensor signals recorded with rhEPO and uhEPO respectively: S(rhEpo)/S(uhEpo) = 0.33.

3.3 *Evaluation of Possible Interfering Compounds Affecting Sensor Selectivity*

Eccipients present in standard solution of rhEPO and uhEPO were tested: NaCl (1,414 ppm) and urea (30 ppm) at physiological levels in buffer (PBS 20 mM, pH = 7.4). No effect on the sensorgram was found. It may be considered uhEPO commercially available originates from anemic patients with anchilostoma infection; containing high proteins levels (72% dosed by *Lowry*) potentially acting as

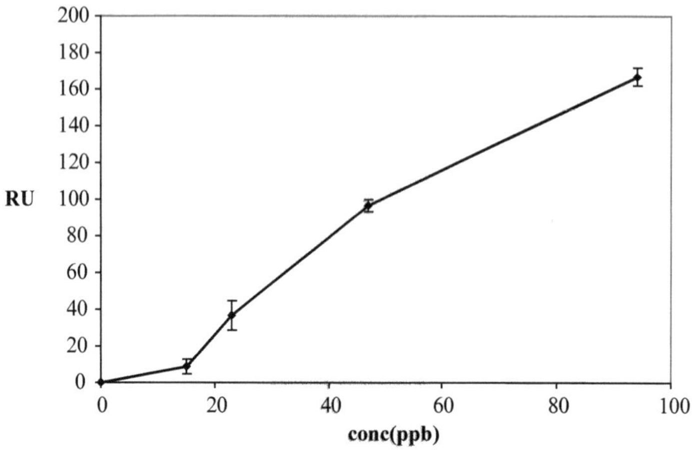

Fig. 2.1 Calibration in buffer, with mAb287 immobilized onto Fc1

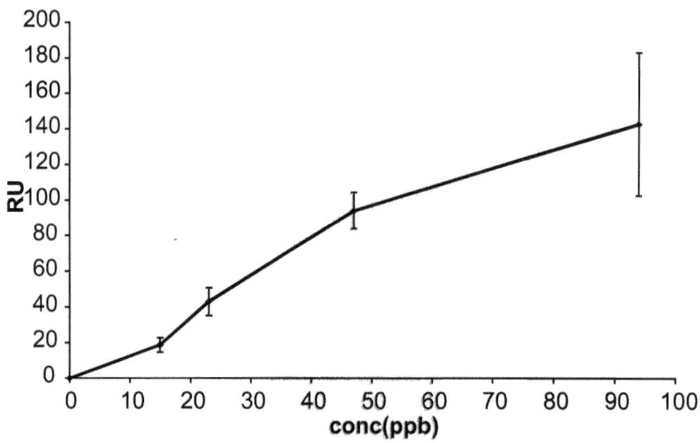

Fig. 2.2 Calibration in urine (1:1 in buffer), with mAb287 immobilized onto Fc1

interfering source (source: uhEPO datasheet). For this reason a high concentration of human serum albumin (HSA) 50 ppm was tested and no interference was found.

Finally the sensor was tested on urine samples selecting a sample dilution: different ratio urine/PBS 20 mM (pH = 7.4) tested alone or spiked with rhEPO.

Samples were filtered in 0.2 μm membrane and heated up to 80°C (3 min), then diluted in buffer PBS 20 mM (1:1) (Fig. 2.2).

The found detection limit (DL) was 23 ppb in spiked urine samples. The measurement reproducibility was low and evaluated as CV% 20%.

Acknowledgments Maria Minunni would like to thank World Antidoping Agengy (WADA) for financial support, within the "Scientific Research Grant –2010 "Detection of Hepcidin as a new Biomarker of Erythropoiesis Stimulators Abuse: A Pilot Study".

References

1. Recny MA, Scoble HA, Kim Y (1987) Structural characterization of natural human urinary and recombinant DNA-derived erythropoietin. Identification of des-arginine 166 erythropoietin. J Biol Chem 262:17156–17163
2. Lasné F, De Ceaurriz J (2000) Nature 405:635
3. Ayotte C, Gmeiner G, Lasnè F, Recombinant erythropoietin in urine. Saugy M, Reichel C, Pascual J (2009) Harmonization of method for the identification of recombinant erythropoietins (i.e. Epoetins) and analogues (e.g. Darbepoetin alfa and methoxypolyethylene glycolepoetin beta). WADA technical document TD2009EPO
4. Bulukin E, Meucci V, Pretti C, Minunni M, Intorre L, Soldani G, Mascini M (2007) An optical immunosensor for rapid vitellogenin detection in plasma from carp (Cyprinus carpio) Talanta 72:785–790

Chapter 3
The Potential Affibodies in New Cancer Marker Immunosensors

Hoda Ilkhani, Marco Mascini, and Giovanna Marrazza

In this paper, a simple and sensitive immunoassay for a tumor marker human epidermal growth factor receptor 2 (HER2) detection is presented. The immunoassay is based on a sandwich format in which modified magnetic microparticles were coupled to multiplexed electrochemical detection platform. The antibody modified beads are used to capture the protein from the sample solution and the sandwich assay is performed by adding Anti-ErbB2 affibody labeled with biotin. The enzyme alkaline phosphatase (AP) conjugated with streptavidin and its substrate (α-naphthyl-phosphate) were then used for the electrochemical detection by a multiplexed electrochemical platform. The immunosensor developed promise to be a sensitive, multiplexed tool for fast and easy HER2 multianalysis.

1 Introduction

The aim of the present work is the investigation of screening devices for the detection of Human Epidermal Growth Factor Receptor 2 (HER2) cancer marker by using antibody modified magnetic microparticles coupled to multiplexed electrochemical detection platform. HER2 also known as ErbB2 is a 185 kDa protein with an intracellular tyrosine kinase domain and an extracellular ligand binding domain [1, 2].

H. Ilkhani
Department of Chemistry, University of Guilan, Rasht, Iran

Center of Excellence in Electrochemistry, University of Tehran, Tehran, Iran

M. Mascini • G. Marrazza (✉)
Dipartimento di Chimica "Ugo Schiff", Università degli Studi di Firenze, Sesto Fiorentino, Firenze, Italy
e-mail: giovanna.marrazza@unifi.it

A. D'Amico et al. (eds.), *Sensors and Microsystems: AISEM 2011 Proceedings*, Lecture Notes in Electrical Engineering 109, DOI 10.1007/978-1-4614-0935-9_3,

In order to improve the performances of immunoassay sandwich format, the secondary antibody was replaced with affibody molecules. Affibodies are a small robust three helical peptide, made up of only 58 amino acids derived from staphylococcal protein A. Affibody molecules bind human epidermal receptor type 2 (HER2) with high affinity and selectivity [3, 4].

The transducers used for the assay were screen printed eight-sensor array, based on eight graphite working electrodes each one with its own silver pseudo-reference electrode and graphite counter electrode.

The novelty of the proposed procedure was the combination of antibody modified magnetic microparticles coupled with affibody molecules and electrochemical platforms as a first step for cancer biomarker multianalysis.

2 Materials and Methods

Dynabeads@ paramagnetic beads, coated with protein A was provided by invitrogen (Milan, Italy). The monoclonal anti human ErbB2 antibody was bought from R&D Systems, from Space srl (Milan, Italy). Anti-ErbB2 affibody@ (biotin) was purchased from Abcam (Cambridge, England). streptavidin-alkaline phosphatase conjugated and α-naphtyl phosphate, were from Sigma (Milan, Italy).

Electrochemical measurements were performed using a PalmSens instrument. All measurements were carried out at room temperature by using Differential Pulse Voltammetry (DPV) with the following parameters: range potential from +0.05 to +0.6 V, step potential 7 mV, potential of pulse 70 mV, time of pulse 0.1 s. Eight-sensors strip coupled with a specially-designed methacrylate well box compatible with the standard eight-channel multi pipette. The sensors array can be applied with the PalmSens CH8 multiplexer configuration. Moreover, for magnetic bio-assays, eight single magnets can be placed at the bottom of the well, thus allowing to concentrate the beads onto each working electrode of the array.

The sample mixer with 12-tube mixing wheels and the magnets were purchased from Dynal Biotech (Milan, Italy).

3 Procedure

A sandwich assay for HER2 detection was developed. This protocol relies on attaching 400 μL of monoclonal anti-human ErbB2 antibody (primary antibody) on 100 μL paramagnetic beads coated with protein A, and blocking free protein A sites with dried milk solution (5% w/v). After 30 min, the tube was positioned on a magnetic block to allow the precipitation of the beads on the bottom of the test tube; the supernatant was then removed and the beads were washed twice with PBS pH 7.4 containing 0.02% Tween 20. The antibody-coated beads could also be prepared in advance and kept at +4°C for several days. The sandwich assay was developed to capture and detect of different concentration of HER2 (200 μL) in 20 mM Tris pH 7.4, 150 mM NaCl. After antigen incubation, the beads were suspended in 5 mg/L biotinylated Anti-ErbB2 affibody (250 μL) for 45 min, in PBS buffer and then were interacted with 500 μL streptavidin-alkaline phosphatase conjugated (2.23 U/mL) for 10 min. Control experiments were performed utilizing the basic procedure described above, expect that the antigen was removed.

10 μL of beads suspension were deposited onto the surface of each working electrode of the array, and kept in its position through the magnet holding block. Thus, each well of the array was filled with 60 μL of a solution containing 1 mg/mL of α-naphthyl phosphate prepared in buffer and was allowed to incubate for 6 min in order to generate sufficient product prior to electrochemical detection of the enzymatically liberated α-naphthol. After incubation time, DPV measurement was left to start. DPV was carried out sequentially for each channel.

All the steps were performed at room temperature, the washing and coating phases were carried out under delicate stirring in the sample mixer and 0.05% of tween 20 was added to all of buffer that used in washing steps.

4 Results and Discussion

The sandwich assay was optimized by several parameters such as concentration of affibody molecules and primary antibody, the kind and incubation time of blocking agent and the kind of buffer. The Table 3.1 shows the results for concentration of affibody and primary antibody, optimization. The optimization was performed by testing the blank and 10 ng/mL solution of HER2. When primary antibody solutions were tested, the affibody solution of 5 mg/L was used. The antibody solution of 50 mg/L was chosen in the following experiments.

The optimized conditions for the immunoassay were: 5 mg/L affibody solution for 45 min, 0.5% w/v dried milk as blocking agent for 30 min, antigen incubation 1 h. The sandwich assay was performed using HER2 standard solutions in the concentration range 5–20 ng/mL (Fig. 3.1). The assay was repeated in order to evaluate the reproducibility; at this purpose, three repetitions of each standard solution were carried out. The limit of detection (LOD) of the assay was evaluated as minimum detectable concentration and was calculated by the evaluation of the average response of the blank plus three times the standard deviation; in this case the recorded blank signal was 0.95 ± 0.11 μA, leading to a LOD of 1.8 ng/mL.

Table 3.1 Optimization of affibody and primary antibody concentration

d	HER2 (0 ng/mL)	HER2 (10 ng/mL)
Primary Ab concentration: 50 mg/L	4.62 ± 0.36 μA	6.96 ± 0.53 μA
Primary Ab concentration: 100 mg/L	3.96 ± 0.23 μA	6.07 ± 0.35 μA
Affibody concentration: 1 mg/L	1.52 ± 0.16 μA	1.31 ± 0.03 μA
Affibody concentration: 5 mg/L	1.52 ± 0.09 μA	1.94 ± 0.04 μA

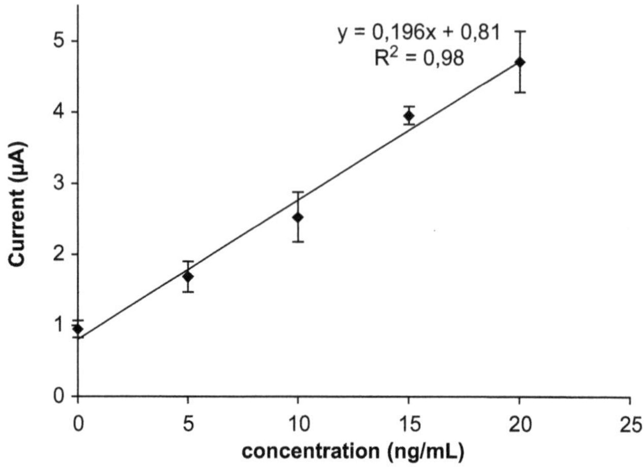

Fig. 3.1 Calibration curve for HER2. The points correspond to the current ± S.D. calculated for three measurements

5 Conclusions

A simple and sensitive approach for HER2 detection in this work was presented, by using antibody modified paramagnetic microparticles coupled to affibody molecules. The paramagnetic microparticles were used as solid phase to build up a sandwich immunoassay, using alkaline phosphatase as enzyme label and signal amplification. Then, an 8-sensors screen-printed array platform was used, blocking the modified beads on working electrodes by means of suitable magnets to carry on electrochemical detection. A good sensitivity and reproducibility were obtained for HER2 detection, with a linear response which matches the request of clinical needs.

References

1. Rubin L, Yarden Y (2001) The basic biology of HER2 Ann. Oncol 12(Suppl 1):S3–S8
2. Coussens L, Yang-Feng TL, Liao Y-C, Chen E, Gray A, McGrath J, Seeburg PH, Libermann TA, Schlessinger J, Francke U, Levinson A, Ullrich A (1985) Tyrosine kinase receptor with extensive homology to EGF receptor shares chromosomal location with neu oncogene. Science 230(4730):1132–1139
3. Nygren PA (2008) Alternative binding proteins: affibody binding proteins developed from a small three-helix bundle scaffold. FEBS J 275:2668–2676
4. Orlova A, Magnusson M, Eriksson TL, Nilsson M, Larsson B, Hoiden-Guthenberg I, Widstrom C, Carlsson J, Tolmachev V, Stahl S, Nilsson FY (2006) Tumor imaging using a picomolar affinity HER2 binding affibody molecule. Cancer Ref 66:4339–4348

Chapter 4
Development of Immunosensor Operating in Organic Mixture for Analysis of Triazinic Pesticides in Olive Oil

Mauro Tomassetti, Elisabetta Martini, and Luigi Campanella

A new immunosensor for triazinic pesticides analysis in hydrophobic matrix such as olive oil and operating in organic mixtures, that is, chloroform-hexane 50% V/V, was developed, analytically characterized and employed in the presence of the oil phase, obtaining good results.

1 Introduction

The quantitative determination of chemical species or various different matrixes that are scarcely soluble or completely insoluble in aqueous solutions has always posed a serious problem in chemical analysis, which has only been partially solved by such techniques as gas chromatography or head space analysis. Biosensor analysis has recently made a substantial contribution to solving this problem through the development of OPEEs, i.e. enzymatic electrodes capable of operating in organic solvents. One classical example is that of inhibition OPEEs to analyze different types of pesticides that are relatively insoluble in aqueous solution, in the development of which also our team has recently been involved [1–3]. The drawback consists in the fact that, rightly or wrongly, it is often complained that inhibition biosensors are relatively unselective, also versus pesticides belonging to different phytopharmaceutical classes. It is a known fact that immunosensors are the most selective biosensors, and our team has recently fabricated immunosensors for triazinic pesticides determination, although operating only in aqueous solution [4]. However, as in the case of enzymatic sensors a few years ago, the greater knowledge acquired in recent times of the, often widely differing, effects exerted by different organic solvents on the protein compounds dissolved in them, has now

M. Tomassetti (✉) • E. Martini • L. Campanella
Department of Chemistry, University of Rome "La Sapienza", Rome, Italy
e-mail: mauro.tomassetti@uniroma1.it

A. D'Amico et al. (eds.), *Sensors and Microsystems: AISEM 2011 Proceedings*, Lecture Notes in Electrical Engineering 109, DOI 10.1007/978-1-4614-0935-9_4,

encouraged several authors to test the possibility of developing antibody-based or similar methods using different solvents or mixtures of organic solvents, mainly water-alcohol mixtures. New immunosensors may be denoted as OPIEs (Organic Phase Immuno Electrodes). With this in mind we undertook a research programme aimed at developing an immunosensor for triazinic pesticides analysis in a hydrophobic matrix such as olive oil operating in (non alcoholic) mixtures of organic solvents, namely chloroform-hexane 50% V/V mixture, in other words, in a mixture that had previously proved to be particularly suitable when enzymatic OPEEs were being developed.

2 Method

To this end, an immunosensor for atrazine was tested in which a hydrogen peroxide electrode was used (see Fig. 4.1) as transducer and peroxidase as marker. In this case the competition process amply described in a previous paper [4] took place in

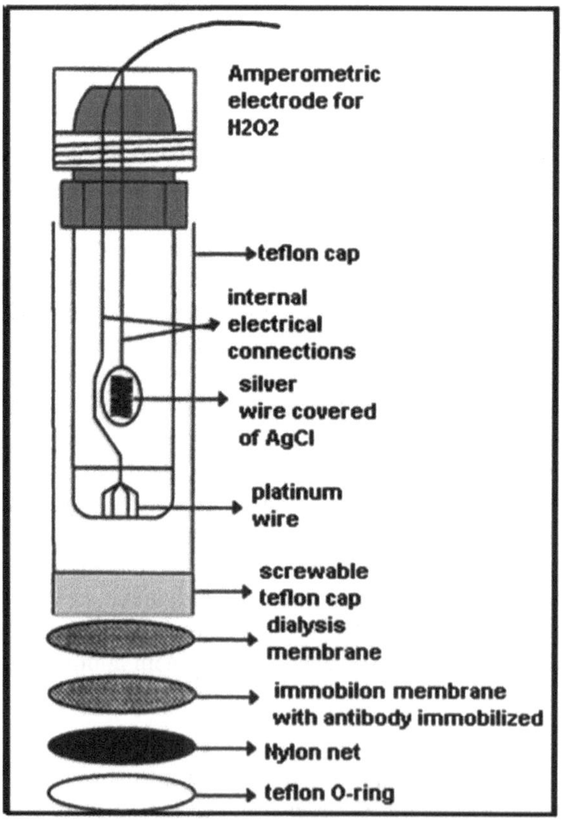

Fig. 4.1 Organic phase immuno electrode (OPIE)

the chloroform-hexane mixture, as stated above, while the subsequent enzymatic measure was performed in a buffered aqueous solution.

3 Results and Discussion

A linear response of between about 0.05 and 5 μmol L^{-1} was obtained versus atrazine for the new OPIE. In Fig.4.2a the behavior of the immunosensor response as a function of growing atrazine concentration, using an Immobilon membrane for antibody immobilization and an amperometric electrode for H_2O_2 as transducer, is shown. In this case the competition was carried out in chloroform – n-hexane (50% V/V) mixture and **in the absence of oil**, while the enzymatic measurement was performed in phosphate buffer.

In Fig. 4.2b the corresponding calibration curve and confidence interval are shown for atrazine determination using a semilogarithmic scale. Conversely, in Fig. 4.3a the behavior of the immunosensor response as a function of growing atrazine concentration is shown, using an Immobilon membrane for antibody immobilization and an amperometric electrode for H_2O_2 as transducer. Also in this case the competition was carried out in chloroform – n-hexane (50% V/V) mixture, but **in the presence of oil**; in this case too the enzymatic measurement was performed in phosphate buffer.

Lastly in Fig. 4.3b the corresponding calibration curve and confidence interval for the atrazine determination, obtained using a semilogarithmic scale, is shown. In Tables 4.1 and 4.2 the main analytical data found in this later two tests are displayed.

It should also be pointed out that if the sensitivity of the test, in which there is no oil in solution during the competitive process, is compared with the sensitivity of the test performed in the same conditions but with 0.5 mL of oil in solution,

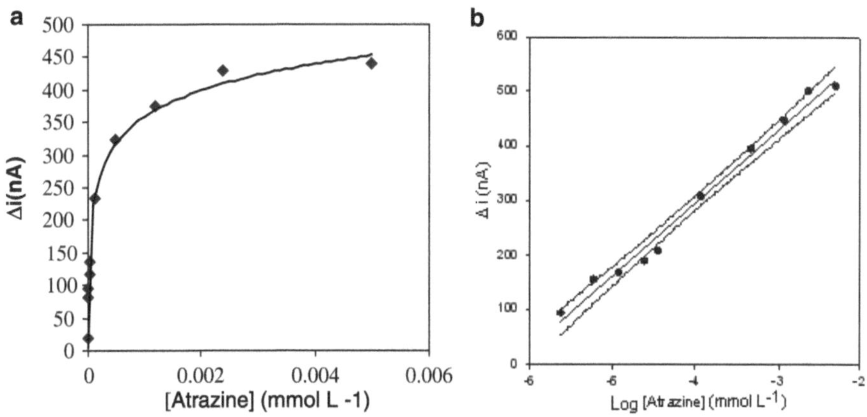

Fig. 4.2 (**a**) Immunosensor response to increasing atrazine concentration, in the absence of oil. (**b**) Corresponding calibration curve

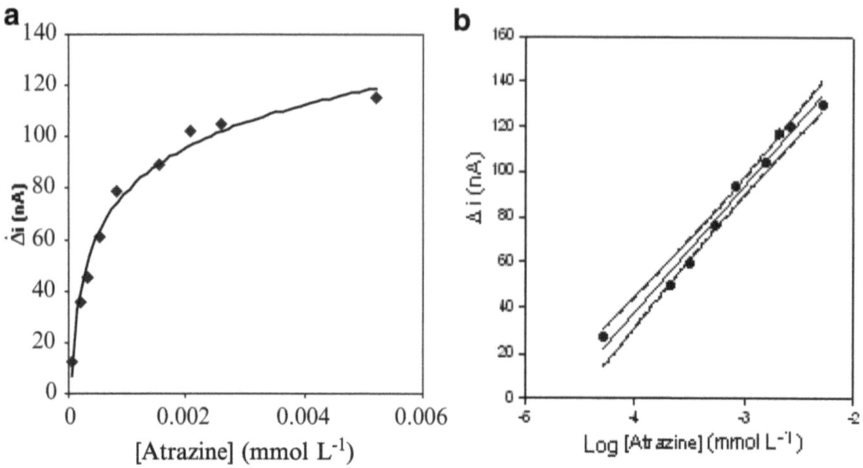

Fig. 4.3 (**a**) Immunosensor response to increasing atrazine concentration in presence of oil. (**b**) Corresponding calibration curve

Table 4.1 Analytical data for atrazine determination using OPIE in the absence of oil

Transducer	Amperometric electrode for H_2O_2
Competition	In chloroform- n-hexane (50% V/V) without oil
Solvent for enzymatic measurement	Phosphate buffer 0.1 mol L^{-1}, pH 7.4
Regression equation ($y = \Delta i$, $x = $ mol L^{-1})	$Y = 134.3\ (\pm 2.3) \log X + 761.6\ (\pm 21.6)$
Linear range (mol L^{-1})	2.5×10^{-9} to 5.0×10^{-5}
Correlation coefficient	0.9881
Repeatability of the measurement (as pooled SD%)	6.4
Low detection limit (LOD) (mol L^{-1})	1.7×10^{-9}

Table 4.2 Analytical data for atrazine determination using OPIE, in the presence of oil

Transducer	Amperometric electrode for H_2O_2
Competition	In chloroform- n-hexane (50% V/V) with oil
Solvent for enzymatic measurement	Phosphate buffer 0.1 mol L^{-1}, pH 7.4
Regression equation ($y = \Delta i$, $x = $ mol L^{-1})	$Y = 55.9\ (\pm 1.2) \log X + 246.6\ (\pm 8.7)$
Linear range (mol L^{-1})	5.0×10^{-8} to 5.0×10^{-6}
Correlation coefficient	0.9833
Repeatability of the measurement (as pooled SD%)	6.7
Low detection limit (LOD) (mol L^{-1})	2.6×10^{-8}

it clearly emerges that in the first case method sensitivity proves to be nearly one and a half times greater than in the second case. This shows that the presence of oil in the chloroform-hexane mixture has an appreciable influence on the competitive process and thus on the method's ultimate sensitivity. Moreover, this is understandable as it is reasonable to assume that the presence of oil can increase viscosity of the organic phase, making the immunocomplex formation slower and more difficult. Nevertheless, the formation of the immunocomplex still takes place in an adequate fashion so that the method can still be satisfactorily applied. Indeed the value of K_{aff}, calculated using the value of IC_{50}, decreases by about only one decade (from 1.48×10^7 L mol^{-1} in the absence of oil, to 1.83×10^6 L mol^{-1} in the presence of oil). These values show that even when the antibody competition occurs in organic solvent, immunocomplex formation takes place more than satisfactorily and allows the immunological method to be developed properly.

4 Conclusions

The results obtained so far have shown that an immunosensor operating in a solvent mixture (such as chloroform-hexane mixture) can be used for the purpose of determining triazinic pesticides in olive oil. Further research is now under way to check if it is possible: (1) to use a Clark type transducer instead of an H_2O_2 electrode for the immunodevice assembly; (2) to perform not only the competition procedure but also the actual final enzymatic reaction in organic solvent instead of in aqueous buffer solution, that is, to carry out the final electro-enzymatic measurement using a classical OPEE operating in organic solvent, or solvent mixture.

Acknowledgments This work was funded by Sapienza University of Rome, "Ateneo Project".

References

1. Campanella L, Lelo D, Martini E, Tomassetti M (2008) Investigation of interfering species in phytodrug analysis using an inhibition Tyrosinase enzyme electrode working both in water and in organic solvent. Anal Lett 41(7):1106
2. Campanella L, Lelo D, Martini E, Tomassetti M (2007) Organophosphorus and carbamate pesticide analysis using an inhibition tyrosinase organic phase enzyme sensor; comparison by butyrylcholinesterase + choline oxidase opee and application to natural waters. Anal Chim Acta 587(1):22
3. Campanella L, Dragone R, Lelo D, Martini E, Tomassetti M (2006) Tyrosinase inhibition organic phase biosensor for triazinic and benzotriazinic pesticide analysis (part two). Anal Bioanal Chem 384:915
4. Campanella L, Lelo D, Martini E, Tomassetti M (2009) Sensors and microsystems. In: D'Amico A, Di Natale C, Martinelli E, Paolesse R (eds) Proceedings of the 13th Italian conference, Rome, 19–21 Feb 2008. World Scientific Printers, Singapore, pp 42–45

Chapter 5
Innovative Electrodes to Control Trace Metal Ionization Used to Treat Pathogens in Water Distribution Systems

Serena Laschi, Ilaria Palchetti, Giovanna Marrazza, and Marco Mascini

The control of hazardous pathogens in water distribution systems, is a priority for health authorities world wide. An estimated 8,000–18,000 people get Legionnaires' disease in the United States each year. Hospitals, hotels, old people's homes, prisons and ships are high risk environments due to the nature of the water distribution system. Treatment is essential, and one of the most effective methods is copper-silver ionization. The positively charged copper and silver ions thus released, form electrostatic bonds with negatively charged sites on bacterial cell walls; this leads to cell lysis and cell death. The amount of copper and silver must remain within a certain range for efficiency, and at the same time remain well below the WHO and other guidelines. High oral intake of copper and silver can result in liver failure and argyria (blue-bluish grey discoloration of the skin) for copper and silver respectively. Recommended values for copper are between 0.3 and 0.5 mg L^{-1} and, for silver, between 0.03 and 0.05 mg L^{-1}.

Currently there is no method available (outside the laboratory) to monitor copper and silver concentrations on-site at a ppb level. The specific aim of this work was to study the electrochemical behaviour of screen-printed graphite electrodes in the determination of silver and copper, with the final purpose of development and construction of mercury-free electrodes to be used in the determination of silver and copper concentrations in water samples by anodic stripping voltammetry.

S. Laschi • I. Palchetti • G. Marrazza (✉)
Dipartimento di Chimica, Università degli Studi di Firenze, Sesto Fiorentino, Italy
e-mail: giovanna.marrazza@unifi.it

M. Mascini
Dipartimento di Chimica, Università degli Studi di Firenze, Sesto Fiorentino, Italy

INBB, Unità di, Firenze, Italy

A. D'Amico et al. (eds.), *Sensors and Microsystems: AISEM 2011 Proceedings*, Lecture Notes in Electrical Engineering 109, DOI 10.1007/978-1-4614-0935-9_5,

1 Materials and Methods

Suprapur grade hydrochloric and nitric acids were purchased from Merck (Milan, Italy). Silver and Copper ICP standard 1 g/L in HNO_3 were purchased from Fluka (Milan, Italy). The water used for preparation of solutions was from a Milli-Q System (Millipore S.p.A., Milan, Italy). The electrochemical cell was composed of screen-printed graphite working and counter electrodes. The working electrode had a diameter of 3 mm. A saturated calomel electrode was used as reference electrode. The silver pseudo-reference electrode that is also printed was cut away from the sensor in order to avoid silver contamination. Electrochemical measurements were performed using an Autolab type II PGSTAT with a GPES 4.9 software package (Metrohm). All measurements were carried out at room temperature.

2 Silver Detection

Silver detection was performed using Anodic Stripping Voltammetry (ASV) as electrochemical technique, with the following parameters: Deposition step: −0.5 V for 120 s, Potential scan range: −0.2/+0.3 V, Step potential 10 mV, Amplitude 0.25 V, Frequency 10 Hz, Conditioning step: +0.5 V for 30 s. The medium used was 0.1 M nitric acid solution containing 14 mM KCl.

2.1 Parameters Optimisation and Silver Measurements

Effect of ASV parameters (amplitude, frequency and step potential) were experimentally evaluated. Thus, each of these parameters were varied at once, whereas the others were kept unvaried. Figure 5.1a shows a tridimentional plot in which is reported how the ASV peak changes in correspondence of amplitude and frequency variations. Keeping fixed the amplitude, the peak current increases with the increase of the frequency. By keeping fixed these two parameters, the step potential was also varied in the range 1–10 mV (Fig. 5.1b).

The final optimised conditions for ASV were found to be: Deposition step: −0.5 V for 120 s, Potential scan range: −0.2/+0.3 V, Step potential 2 mV, Amplitude 0.25 V, Frequency 250 Hz, Conditioning step: +0.5 V for 30 s. A calibration curve for Silver was then carried out, and a linear range obtained in the concentration range 0–30 ppb, associated to an average RSD of 12%. A detection limit of 0.55 ppb was also evaluated, demonstrating the high sensitivity of the system.

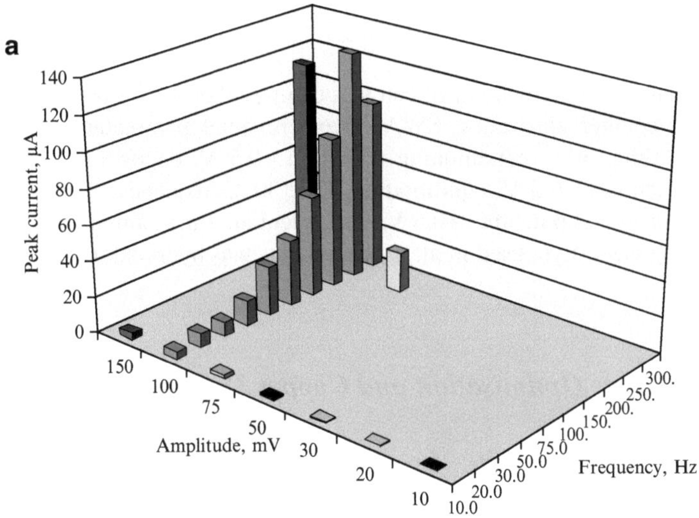

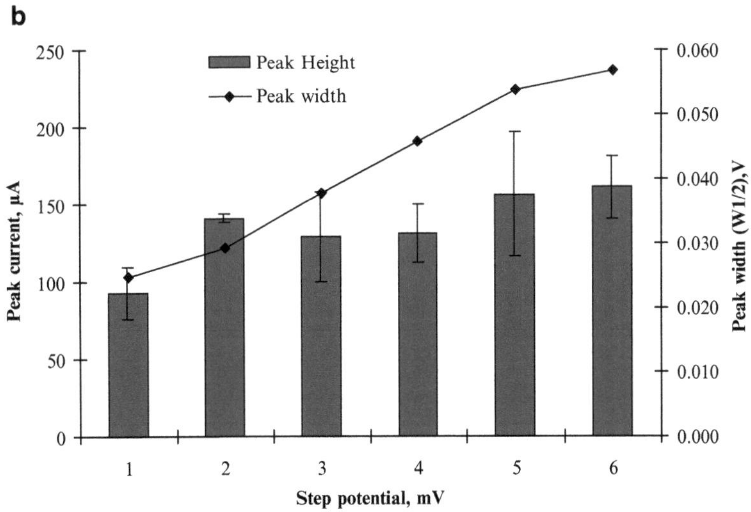

Fig. 5.1 (**a**) Tridimentional plot showing the effect on the peak height of the amplitude and of the frequency in ASV measurements. Conditions used are: deposition step: −0.5 V for 120 s, potential scan range: −0.2/+0.3 V, step potential 10 mV, conditioning step: +0.5 V for 30 s. Medium used: 0.1 M nitric acid solution containing 14 mM KCl. Ag(I) concentration tested: 100 ppb. (**b**) Graph showing the effect on the peak height of step potential, at fixed amplitude and frequency. Conditions used are: deposition step: −0.5 V for 120 s, potential scan range: −0.2/+0.3 V, frequency 250 Hz , amplitude 0.25 V, conditioning step: +0.5 V for 30 s. Medium used: 0.1 M nitric acid solution containing 14 mM KCl. Ag(I) concentration tested: 100 ppb

3 Copper Detection

Also in this case the electrochemical cell was composed of screen-printed graphite working and counter electrodes. SWASV experimental parameters used were: Conditioning time: 30 s, conditioning potential: +0.5 V, deposition time: 15 s, deposition potential: −0.5 V, equilibration time: 30 s, frequency: 250 Hz, initial potential: −0.5 V, end potential: +0.5 V, step potential: 2 mV, amplitude:100 mV. The supporting electrolyte used in all measurements was hydrochloric acid 0.1 M.

3.1 Parameters Optimisation and Copper Detection

Between the different parameters, supporting electrolyte composition and deposition time were optimized. [1, 2]. Different times were used in the deposition step: 15, 30, 60 and 90 s respectively. Then, using the same sensor for consecutive Cu(II) additions, calibration curves were constructed by evaluating the peak shape, peak height and the linearity range.

As it can be seen from Fig. 5.2, the broader linearity range was achieved for the shorter deposition time tested. In addition, also the linearity range is more extended when the deposition time is short, as reported in Table 5.1.

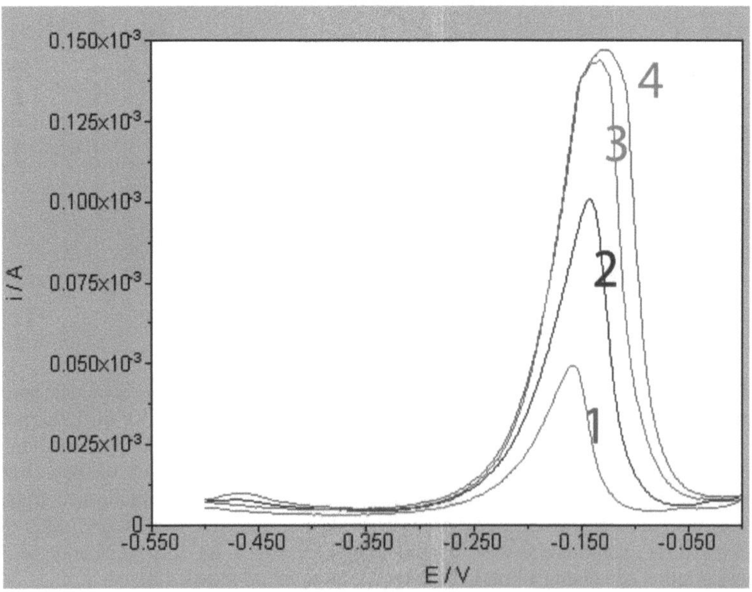

Fig. 5.2 SAWSV peak shapes using (1) 15, (2) 30, (3)60 and (4) 90 s as deposition time respectively. Cu(II) tested concentration 400 ppb. Other SWASV conditions: deposition potential: −0.5 V; stripping: −0.5 V to +0.5 V, Eamp: 100 mV; Estep: 2 mV; f: 250 Hz

Table 5.1 Linearity range obtained for Cu(II) calibration curve performed with different deposition time. Other SWASV conditions: deposition potential: −0.5 V; stripping: −0.5 V to +0.5 V, Eamp: 100 mV; Estep: 2 mV; f: 250 Hz

Deposition time (s)	Linearity range (ppb)
15	up to 1,200
30	up to 700
60	up to 500
90	up to 500

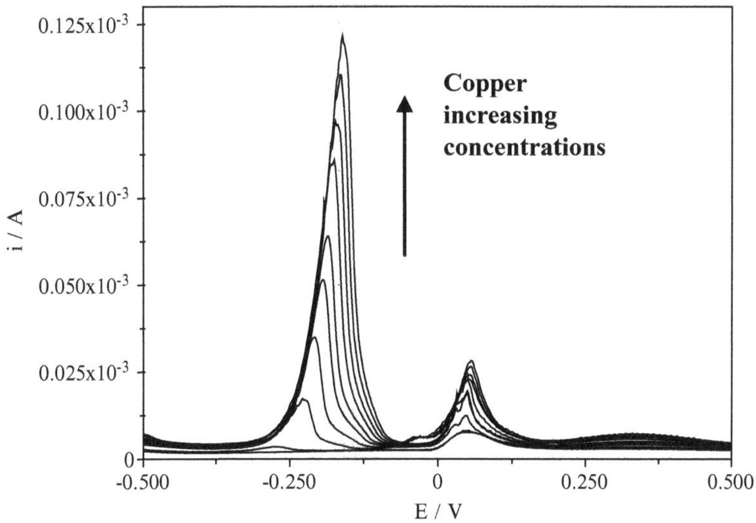

Fig. 5.3 SWASV voltammograms of increasing concentrations of Cu(II)

Finally, calibration experiments were performed using square wave anodic stripping voltammetry (SWASV) at optimised conditions, using the same electrode for all concentrations. Figure 5.3 shows SWASV peaks obtained in the range 0–1,300 ppb.

Experimental results indicate that a very good linearity was found until around 1,200 ppb as Cu(II) concentration, with a high reproducibility (average RSD: 2.3%).

4 Conclusions

It can be affirmed that graphite-based screen-printed sensors were identified as a good electrochemical substrate for Silver and Copper detection in water samples. Measurements cannot be carried on simultaneously because different experimental conditions need to be used, thus experimental conditions were optimised separately. Nevertheless, in both cases the analysis of the sample show interesting analytical features; actually the dynamic range observed was of analytical interest, so that these devices can be considered as a valid instrument for metals detection in water samples.

References

1. Wantz F, Banks CE, Compton RG (2005) Edge plane pyrolytic graphite electrodes for stripping voltammetry: a comparison with other carbon based electrodes electroanalysis 17:651–655
2. Neuhold CG, Wang J, Nascimento VB, Kalcher K (1995) Thick film voltammetric sensors for trace copper based on a cation-exchanger-modified surface. Talanta 42:1791–1798

Chapter 6
High-Sensitive Impedimetric Aptasensor for Detection Ochratoxin A in Food

Gabriela Castillo, Ilaria Lamberti, Lucia Mosiello, and Tibor Hianik

We report a high sensitive biosensor based on DNA aptamers for detection ochratoxin A (OTA). The thiolated DNA aptamers specific to OTA have been immobilized to a surface of gold electrode. The electrochemical impedance spectroscopy (EIS) at presence of redox probe $[Fe(CN)_6]^{-3/-4}$ has been used for determination of charge transfer resistance, R_{ct}, following addition of OTA. We have shown that R_{ct} increased with increasing OTA in a range of 0.1–100 nM. The sensor revealed high sensitivity (the limit of detection was 0.44 nM), selectivity and was regenerable. The sensor was validated in coffee extract.

1 Introduction

Electrochemical biosensors based on DNA aptamers are of growing interest due to their high sensitivity and selectivity [1]. This is particularly due to their high affinity to proteins or to other compounds, which is comparable with that of antibodies. In contrast with antibodies, aptamers are synthesized in vitro by the SELEX procedure [2]. Aptamers can be chemically modified by biotin, thiol or amino groups, which

G. Castillo • T. Hianik (✉)
Department of Nuclear Physics and Biophysics, Comenius University, Bratislava, Slovakia
e-mail: tibor.hianik@fmph.uniba.sk

I. Lamberti
Department of Biology, Universirity of Roma Tre, Roma, Italy

L. Mosiello
ENEA, Italian National Agency for New Technologies, Energy and the Environment, Roma, Italy

A. D'Amico et al. (eds.), *Sensors and Microsystems: AISEM 2011 Proceedings*, Lecture Notes in Electrical Engineering 109, DOI 10.1007/978-1-4614-0935-9_6,

allowing them to be immobilized on various solid supports. This opens new routes for construction of biosensors for practical applications. Recently the DNA aptamer sensitive to mycotoxin – ochratoxin A (OTA) has been developed [3]. OTA belongs to toxical fungal metabolites that can occur in primary food products [4]. Most recently, the European Commission has fixed maximum concentration of OTA in foodstuffs: 3 μg/kg (7.4 nM) for cereal products and 5 μg/kg (12.4 nM) for roasted coffee, respectively (EC No. 1881/2006). The development of fast and efficient method of detection OTA is of high importance. The OTA sensitive biosensors developed so far were based on oxidation of OTA at glassy carbon electrode [5], reduction of horseradish peroxidase [6] or using antibodies immobilized on carbon nanotubes [7]. According to our knowledge the first aptasensor for OTA was reported most recently and utilized electrochemiluminiscence detection [8]. Further the electrochemical aptasensor based on competitive assay (Limit of detection (LOD) 0.27 nM) [9] has been developed. However, in this work the OTA conjugated with alkaline phosphatase have to be used in competitive assay. The direct detection of OTA would be, however, more advantageous for practical applications. In present work we developed the impedimetric aptasensor for detection OTA with comparable sensitivity (LOD 0.44 nM).

2 Materials and Methods

The DNA aptamer selective to OTA was adopted from [3] (aptamer 1.12.2), but contained dT_{15} spacer terminated by SH group: 5′ GAT CGG GTG TGG GTG GCG TAA AGG GAG CAT CGG ACA – dT_{15}- 3′- SH. Aptamers were purchased from Thermo Fisher Scientific (Ulm, Germany) and immobilized at the clean gold electrode (diameter 2 mm) by chemisorption (see [10] for cleaning procedure). Briefly, the electrode was immersed in 5 μM aptamers dissolved in TE buffer (10 mM Tris–HCl + 1 mM EDTA, pH 7.6) and kept 16 h at 4°C. Then it was rinsed by deionised water and immersed in 100 mM mercaptoethanol (Aldrich, USA) for 20 min. The sensor was rinsed several times in binding buffer (10 mM HEPES + 120 mM NaCl + 5 mM KCl + 20 mM $CaCl_2$, pH 7.0) and then incubated during 30 min in OTA (0.1–100 nM) or 100 nM Ochratoxin B (OTB) (Romer Labs, Austria). The sensor has been validated in spiked 10% coffee extract.

Experiments were performed with AUTOLAB PGSTAT12 (Eco Chemie, The Netherlands) in a Teflon cell of a 5 mL volume using three electrode configuration: working gold electrode a Pt wire (auxiliary electrode) and Ag/AgCl reference electrode (CH Instruments, USA). The electrochemical impedance spectroscopy (EIS) was performed with FRA module of AUTOLAB in a frequency range 0.1 Hz–100 kHz (AC voltage 5 mV, DC voltage 0.165 mV).

3 Results and Discussion

EIS is highly sensitive to surface modifications. Using redox probe $[Fe(CN)_6]^{-3/-4}$ it is possible to amplify the detection of ligands by aptamers at surfaces. Because DNA aptamer is negatively charged this anionic probe will repeal from the sensor surface causing increase in charge transfer resistance (R_{ct}). Binding of protein resulted in decrease (lysozyme) or increase (thrombin) the negative charge at the surface which resulted in decrease or increase of R_{ct}, respectively [11, 12]. We showed that this method can be applied also for determination of OTA. The Nyquist plot of the aptasensor at various concentrations of OTA is presented on Fig. 6.1. The plot consists of a semicircle at higher frequencies and of linear part at lower frequencies. The diameter of semicircle is proportional to the R_{ct} value while the linear part corresponds to the diffusion of the probe which is represented by the Warburg impedance. With increased concentration of OTA the diameter of semicircles increases which is evidence of increased R_{ct} values. At experimental condition used OTA is negatively charged [4] and thus its binding to aptamers should increase the negative charge of the surface and induce repealing of redox probe. Therefore R_{ct} value should increase. The plot of relative changes of R_{ct} vs. OTA concentration is presented on Fig. 6.2 It can be seen that at relatively low OTA concentration range 0.1–3 nM sharp increase of OTA took place with subsequent saturation at higher toxin concentrations. The Plot has the shape of Langmuir isotherm [13] and can be described as: $(\Delta R_{ct}/R_{ct0}) = (\Delta R_{ct}/R_{ct0})_{max}\ [c/(K_D + c)]$, where $(\Delta R_{ct}/R_{ct0})_{max}$ is the maximum R_{ct} variation and c is the OTA concentration. The $K_D = 3.62 \pm 2.31$ nM is lower in comparison with that determined by fluorescence method for free aptamers in a solution [3]. This may be evidence of improved affinity of aptamers immobilized at surface. LOD determined for signal to noise level S/N = 3 was 0.44 nM (0.16 ppb), which is sufficient for practical use.

The aptasensor was selective to OTA. No significant changes in R_{ct} were observed at presence of 100 nM OTB. It was also possible to easily regenerate the sensor surface by immersion this for 30 s in 1 mM HCl and by washing in

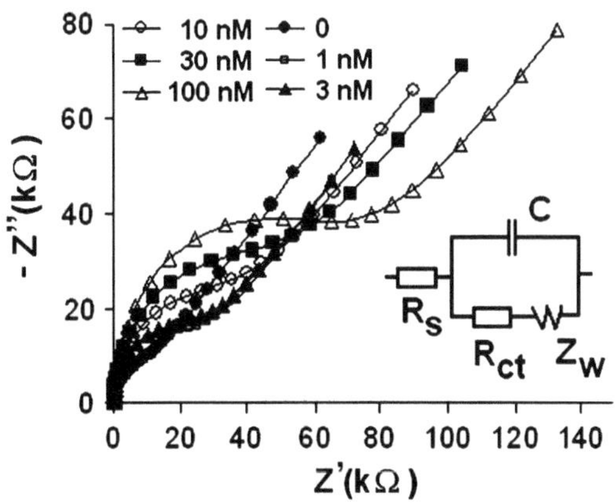

Fig. 6.1 Nyquist plot of the aptasensor with different OTA concentrations at presence of 1 mM $[Fe(CN)_6]^{-3/-4}$. *Inset*: the Randles equivalent circuit: R_S – electrolyte resistance, C – interfacial capacity, R_{ct} – charge transfer resistance and Z_W – Warburg element

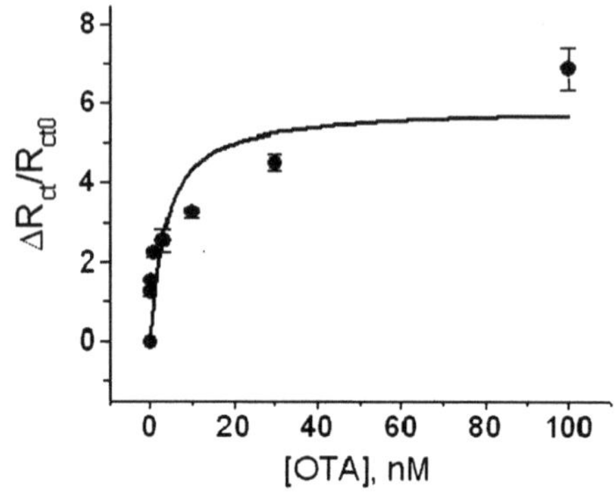

Fig. 6.2 The plot of the relative changes of charge transfer resistance, $\Delta R_{ct}/R_{ct0}$, as a function of OTA concentration. The full line is the fit according to Langmuir isotherm (see the text). Results represent mean ± S.D. obtained at three independently prepared biosensors

Table 6.1 The determination of OTA by aptasensor in spiked coffee. Results represent mean ± S.D. obtained from three independent experiments

Sample	OTA added, ppb	OTA found, ppb	Recovery, %
Coffee	1	0.85 ± 0.03	85
	5	4.40 ± 0.04	87
	10	8.80 ± 0.03	88

deionised water [14]. We also validated the biosensor for determination OTA in spiked coffee. The sensor response to OTA in a coffee was similar to that in a buffer (Table 6.1).

4 Conclusions

The developed aptasensor revealed high sensitivity and selectivity. It can be easily prepared and regenerated and used for determination of OTA in food.

Acknowledgments Authors thank to Agency for Promotion Research and Development (contracts No. APVV-410-10) and VEGA (grant No. 1/0794/10) for financial support.

References

1. Hianik T, Wang J (2009) Electrochemical aptasensors – recent achievements and perspectives. Electroanalysis 21:1223–1235
2. Tuerk C, Gold L (1990) Systematic evolution of ligands by exponential enrichment: RNA ligands to bacteriophage T4 DNA polymerase. Science 249:505–510
3. Cruz-Aguado JA, Penner G (2008) Determination of ochratoxin A with a DNA aptamer. J Agric Food Chem 56:10456–10461
4. El Khoury A, Atoui A (2010) Ochratoxin A: general overview and actual molecular status. Toxins 2:461–493

5. Oliveira SCB, Diculescu VC, Palleschi G, Compagnone D, Oliveira-Brett AM (2007) Electrochemical oxidation of ochratoxin A: at a glassy carbon electrode and in situ evaluation of the interaction with deoxyribonucleic acid using an electrochemical deoxyribonucleic acid – biosensor. Anal Chim Acta 588:283–291
6. Alonso-Lomillo MA, Domínguez-Renedo O, Ferreira-Goncalves L, Arcos-Martínez MJ (2010) Sensitive enzyme-biosensor based on screen-printed electrodes for Ochratoxin A. Biosens Bioelectron 25:1333–1337
7. Kaushik A, Solanki PR, Pandey MK, Kaneto K, Ahmad S, Malhotra BD (2010) Carbon nanotubes chitosan nanobiocomposite for immunosensor. Thin Solid Films 519:1160–1166
8. Wang ZP, Duan N, Hun X, Wu SJ (2010) Electrochemiluminiscent aptamer biosensor for the determination of ochratoxin A at a gold nanoparticles-modified gold electrode using N-(aminobutyl)-N-ethylisoluminol as a luminescent label. Anal Bioanal Chem 398:2125–2132
9. Barthelmebs L, Hayat A, Limiadi AW, Marty J-L, Noguer T (2011) Electrochemical DNA aptamer-based biosensor for OTA detection, using superparamagnetic nanoparticles. Sensor Actuat B 156:932–937. doi:0.1016/j.snb.2011.03.008
10. Hianik T, Ostatna V, Zajacova Z, Stoikova E, Evtugyn G (2005) Detection of aptamer–protein interactions using QCM and electrochemical indicator methods. Bioorg Med Chem Lett 15:291–295
11. Rodriguez MC, Kawde A-N, Wang J (2005) Aptamer biosensor for label-free impedance spectroscopy detection of proteins on recognition-induced switching of the surface charge. Chem Commun 34:4267–4269
12. Lee JA, Hwang S, Kwak J, Park SI, Lee K-C (2008) An electrochemical impedance biosensor with aptamer-modified pyrolyzed carbon electrode for label-free protein detection. Sensor Actuat B 129:372–379
13. Wang J (2006) Analytical electrochemistry. Wiley-VCH, New York
14. Tombeli S, Minunni M, Mascini M (2005) Piezoelectric biosensors: strategies for coupling nucleic acids to piezoelectric devices. Methods 37:48–56

Chapter 7
Introduction of an Electrochemical Genosensor for Detection of P53 Gene Via Sandwich Hybridization Method

Ezat Hamidi-Asl, Ilaria Palchetti, and Marco Mascini

Sequence-specific detection of DNA provides the basis for detection of a wide variety of infectious and inherent disease. Electrochemical hybridization biosensors for the detection of DNA sequences reduce the assay time and simplify medical analysis. In this work, we introduce a biosensor for investigation of DNA hybridization related to p53 gene corresponding oligonucleotide. Due to the high importance of p53 gene detection in various fields of medicine and biology, we attempted to prepare a rapid and sensitive enzyme-linked electrochemical genosensor using screen printed electrodes (SPE). Hybridization was performed on streptavidin coated paramagnetic micro-beads functionalized with a biotinylated capture probe. The complementary sequence was then recognized via sandwich hybridization with a capture probe and a biotinylated signaling probe. After labeling the biotinylated hybrid with a streptavidin–alkaline phosphatase conjugate, the particles were introduced onto a disposable SPE. The modified particles were trapped with a magnet onto the sensor surface. Then a known volume of a solution containing the enzymatic substrate, α-naphthyl phosphate, was added on the SPE surfaces. The oxidation signal of the enzymatically produced α-naphthol was investigated by differential pulse voltammetry. The genosensor response varied linearly with the oligonucleotide target concentration between 0.05 nM and 2.0 nM. The estimated detection limit was 0.03 nM.

E. Hamidi-Asl
Electroanalytical Chemistry Research Laboratory, Department of Analytical Chemistry, University of Mazandran, Babolsar, Iran

I. Palchetti (✉) • M. Mascini
Dipartimento di Chimica, Università degli Studi di Firenze, Sesto Fiorentino, Italy
e-mail: ilaria.palchetti@unifi.it

A. D'Amico et al. (eds.), *Sensors and Microsystems: AISEM 2011 Proceedings*, Lecture Notes in Electrical Engineering 109, DOI 10.1007/978-1-4614-0935-9_7,

1 Introduction

Sequence-specific detection of DNA targets has become increasingly important in molecular diagnostics. In these years, electrochemical DNA biosensors are widely applied for the rapid and inexpensive diagnosis of biological analysis due to their prominent advantages. They are simple, portable, rapid, precise, sensitive and inexpensive. A typical electrochemical DNA biosensor is made of a solid electrode with immobilized short single-stranded DNA probe on it and electroactive hybridization indicators. The hybridization between DNA probes with complementary sequences influences the performance of electrochemical DNA biosensors. This affection depends on the selection of the probes and the conditions of hybridization. Therefore, design the DNA electrochemical biosensor depends on the kind of hybridization and generation of electrochemical signal [1].

The gene mutation takes place in around 60% in all humans' lifetimes [2]. Therefore, introduction of new methods for mutation detection is very important. Different methods are used to analyse the mutation status of individual tumours [3]. It has been estimated that in 50% of human tumours, p53 gene is mutated [4]. This demonstrates the important role of the p53 pathway in regulation of cell growth and survival. The p53 status of a tumour may have a strong influence on sensitivity to commonly used anticancer drugs and radiotherapy. Therefore, p53 is an extraordinary clinical marker and new therapeutic target. The tumour suppressor p53 has been signified in a expanded number of biological processes, including cell cycle arrest, senescence, apoptosis, autophagy, metabolism, and aging. Activation of p53 in response to oncogenic stress deletes nascent tumour cells by apoptosis or senescence. Improved studying of the p53 pathway may lead to better diagnosis and treatment of cancer in the future [5].

In this paper, we report a method for detection p53 gene corresponding oligonucleotide. In the current work, streptavidin-coated paramagnetic micro-beads were modified with a biotinylated capture probe. The complementary sequence corresponding to the short sequence of p53 gene was then recognized via sandwich hybridization with a capture probe and a biotinylated signaling probe. After labeling the biotinylated hybrid with a streptavidin–alkaline phosphatase conjugate, the particles were introduced in a disposable screen printed electrodes (SPEs). To perform electrochemical measurements SPEs were kept horizontally and a magnetic holding block was placed on the bottom part of the electrode, to better localize the beads onto the working surface. Then, the modified beads were incubated with the enzymatic substrate, α-naphthyl phosphate, solution. α-naphthol was enzymatically produced and oxidised around 0.2 V vs. screen- printed silver pseudo- reference electrode using differential pulse voltammetry. The α-naphthol oxidation peak was taken as the analytical signal.

2 Materials and Methods

2.1 Reagents

Tween 20, Streptavidin–alkaline phosphatase, α-naphthyl phosphate, bovine serum albumin (BSA), magnesium chloride and diethanolamine were obtained from Sigma–Aldrich. Disodium hydrogenphosphate, ethylendiamine tetra acetic acid (EDTA) and potassium chloride were purchased fromMerck. MilliQ water was used throughout this work. Streptavidin-coated paramagnetic beads were purchased from Invitrogen (Dynabeads MyOne Streptavidin C1, Invitrogen Dynal AS, Oslo, Norway). Synthetic oligonucleotides were obtained from MWG Biotech. The sequences of synthetic oligonucleotides are below:

Capture probe (15mer): biotin-5′-AGT TCT CCA TCC CCA-3′, Signaling probe (15mer): 5′-GGA GAG ATG CTG AGG-biotin-3′, target (30 mer): CCT CAG CAT CTC TCC TGG GGA TGG AGA ACT.

2.2 Disposable Screen-Printed Electrodes

Screen-printed electrodes (SPEs) are made of a carbon working electrode, a carbon counter electrode and a silver pseudo reference electrode. Materials and procedures to screen-print the transducers are described elsewhere [6]. Electrochemical measurements were done with μAutolab type II PGSTAT with a GPES 4.9 software package (Metrohm, Rome, Italy). All potentials were referred to the Ag/AgCl pseudo-reference electrode. All experiments were performed at room temperature. During electrochemical measurement, SPEs were kept horizontally and a magnet was placed on the bottom of the electrode. Then, enzymatic substrate was added on the SPE surface to produce α-naphthol.

2.3 Hybridization Procedure

Sandwich-like format was used for hybridization experiments. Synthetic target was diluted with a solution 0.2 μM of biotinylated signaling probe in phosphate buffer 0.5 M pH 7. For every assay 20 μL of probe-modified beads were employed. Using the magnetic particle concentrator, the buffer was removed carefully and then the beads were incubated with 50 μL of the hybridization solution for 30 min. After hybridization, the beads were washed three times with 100 μL of DEA buffer (0.1 M diethanolamine, 1 mM $MgCl_2$, 0.1 M KCl, tween 20, 0.05%; pH 9.6), to remove nonspecifically adsorbed sequences.

2.4 Labeling with Alkaline Phosphatase

Following the hybridization and three times washing steps, the beads were immersed in 50 μL of a solution containing 0.75 U/mL of the streptavidin–alkaline phosphatase and 10 mg/mL of BSA (blocking agent) in DEA buffer. After 20 min, beads were washed three times with 100 μL of DEA buffer.

2.5 Electrochemical Detection

In the next stage, the beads were immersed in 50 μL of DEA buffer. The electrochemical measurements were carried out by placing a magnetic particle concentrator under the SPEs; 10 μL of enzyme-labeled bead suspension was deposited on the working electrode surface. Then, 60 μL of 1 mg/mL α -naphthyl phosphate in DEA buffer was added to the SPEs. After 5 min, the electrochemical signal of the enzymatically produced α-naphthol was measured by DPV (modulation time 0.05 s; interval time 0.15 s; step potential 5 mV; modulation amplitude 70 mV; potential scan from 0 to +0.6 V). The height of α-naphthol oxidation peak was investigated as the analytical signal. Reported results are the average of at least three measurements and the error bars correspond to the standard deviation.

3 Results and Discussion

Initial experiments were done using SPEs in order to optimize some parameters (i.e., modification of beads, assay times and procedures). To investigate the analytical performances of the genosensor with synthetic oligonucleotides, a calibration

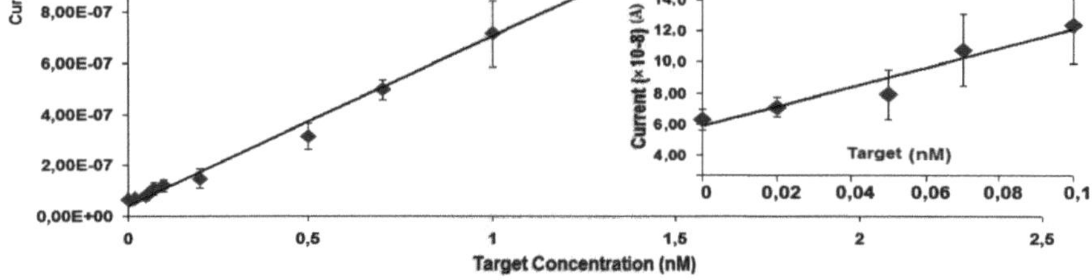

Fig. 7.1 Calibration plot for different concentrations of p53 DNA target oligonucleotide

experiment was designed. The DNA sensor response varied linearly ($r^2 = 0.99$) with the target concentration over two orders of magnitude, between 0.05 nmol/L and 2 nmol/L (Fig. 7.1). Within the linear analytical range, the sensitivity was 7×10^{-7} (A/nM), with an estimated detection limit of 0.03 nmol/L. A calibration experiment is shown in the Fig. 7.1.

4 Conclusions

In this work a rapid and sensitive enzyme-linked electrochemical genomagnetic assay by a sandwich hybridization system was developed. Using of paramagnetic beads allowed the possibility to measure nM level of DNA sequences, with high reproducibility.

References

1. Wang J, Kawde A, Erdem A, Salzar M (2001) Magnetic bead-based label-free electrochemical detection of DNA hybridization. Analyst 126:2020–2024
2. Jiang T, Minunni M, Wilson P, Zhang J, Turner APF, Mascini M (2005) Detection of TP53 mutation using a portable surface plasmon resonance DNA-based biosensor. Biosens Bioelectron 20:1939–1945
3. Nollau P, Wagener C (1997) Methods for detection of point mutations: performance and quality assessment. Clin Chem 43:1114–1128
4. Soussi T, Wiman KG (2007) Shaping genetic alterations in human cancer: the p53 mutation paradigm. Cancer Cell 12:303–312
5. Farnebo M, Bykov VJN, Wiman KG (2010) The p53 tumour suppressor: a master regulator of diverse cellular processes and therapeutic target in cancer. Biochem Biophys Res Commun 396:85–89
6. Laschi S, Palchetti I, Marrazza G, Mascini M (2006) Development of disposable low density screen-printed electrode arrays for simultaneous electrochemical measurements of the hybridisation reaction. J Electroanal Chem 593:211–218

Chapter 8
Peptide Modified Gold Nanoparticles for the Detection of Food Aromas

Giuseppe C. Fusella, D. Compagnone, Caterina I. Saulle, R. Paolesse, and C. Di Natale

In this study we attempted to add new functionalities to piezoelectric sensors based e-nose by modification with gold nanoparticles (GNPs) bearing short peptide moieties. The modified piezoelectric sensors have been tested for model water solutions of aromas as isopentyl acetate, etylacetate, cis-3-hexen-1-ol and terpinen-4-ol. Principal component analysis of the data, compared with data obtained using a porphyrin based electronic nose demonstrated that this approach can be useful for the development of a new type of piezoelectric sensors for e-nose.

1 Introduction

It is well known that food aromas are very important for customers satisfaction. The addition of aromatic compounds and/or the release of them during processing steps needs to be monitored for the control of the quality of the final product. Odours and food aromas results from a complex equilibrium between the amount different volatile species having different chemical nature; they are released in the vapor phase depending on the physico-chemical structure of food. Direct measurement of aromas via e-noses represent a viable alternative to classical gas-chromatographic

G.C. Fusella • D. Compagnone (✉) • C.I. Saulle
Dipartimento di Scienze degli Alimenti, Universita´ degli studi di Teramo,
Mosciano Sant'Angelo (TE), Italy
e-mail: dcompagnone@unite.it

R. Paolesse
Dipartimento di Scienze e Tecnologie Chimiche, Universita´degli
studi di Roma Tor Vergata, Rome, Italy

C. Di Natale
Dipartimento di Ingegneria Elettronica, Universita´degli studi di Roma Tor Vergata,
Rome, Italy

A. D'Amico et al. (eds.), *Sensors and Microsystems: AISEM 2011 Proceedings*,
Lecture Notes in Electrical Engineering 109, DOI 10.1007/978-1-4614-0935-9_8,

analysis, particularly for classification and for process monitoring. In this respect, QCM based e-noses, working at low temperature, give the chance to chemically modify the sensors surfaces in order to improve functionalities of the measurement [1]. In this work we attempted to use gold nanoparticles bearing aminoacids and short peptides to modify the surface of the sensors. The use of nanostructured material, such as gold nanoparticles (GNP) is expected to add new functionalties to the sensors because of the larger surface interaction, and the large number of possible modifications using peptides [2]. In this preliminary work we tested five different GNP based QCM sensors for the detection of typical food aromas and compared the data with porphyrin based QCM-sensors.

2 Materials and Methods

2.1 GNP Synthesis Process

All chemicals were purchased from Sigma-Aldrich. GNP were prepared by the sodium borohydride reduction method [3]. In a typical experiment, 100 mL aqueous solution of tetrachloroauric acid (10^{-4} M) was reduced by 0.01 g of $NaBH_4$ at room temperature resulting in the formation of ruby-red gold hydrosol containing gold nanoparticle with diameter of about 7 nm. GNPs were modified using 10^{-4} M aqueous solution of cysteine (CYS), glutathion (GSH), γ-glutammylcystein (γ-GLU-CYS) and cysteinylglycine (CYS-GLY). GNP have been characterized using TEM, VIS spectroscopy and electrochemistry. The different GNP solution were used to modify 20 MHz QCM, by drop casting of 50 μL on each side of the sensor crystal.

2.2 Electronic Nose Analysis

The electronic nose used was developed in University of Tor Vergata in the Department of Electronic Engineering. The system consisted in eight different channels to measure variation frequency and used porphyrin based sensors as previously reported [1]. A sampling protocol of model water solutions of aromas (isopentyl acetate, etylacetate, cis-3-hexen-1-ol and terpinen-4-ol) headspace was designed. A 1,000 ml flask was filled with 10 ml of water solution, sealed and kept at room temperature. The headspace was aspirated into the electronic nose sensor cell by a constant flow of air.

3 Results and Discussion

The UV–vis spectra confirmed the formation of GNP by the characteristic plasmon resonance band of GNP at 512 nm. The shift of the plasmon band confirm the correct modification as reported by Aryal et al. [3]. GNP, GNP-CYS, GNP-GSH, GNP-(CYS-GLY) and GNP-(γ-GLU-CYS) were used to obtain QCMs with new active surface area and compared with porphyrin modified QCMs. In the experimental conditions tested, the changes in frequency for each of the sensors for the four different aromas typically found in foods (cis-3-hexen-1-ol, terpinen-4-ol, ethyl acetate and isopenthyl acetate) are reported in Fig. 8.1. RSD was below 15% for each sensor. Differences in the selectivity patterns among peptide derived GNPs and porphyrin based sensors are evident and appear promising for the development of gas sensor based on the new functionalities.

Figure 8.2 reports PCA score plots obtained from the datasets. The PCA model explains the 81% of the variance for the seven porphyrin based sensors and 99% of variance for the five GNP sensors. Both set of sensors are able to discriminate among the different aromas in the model solutions.

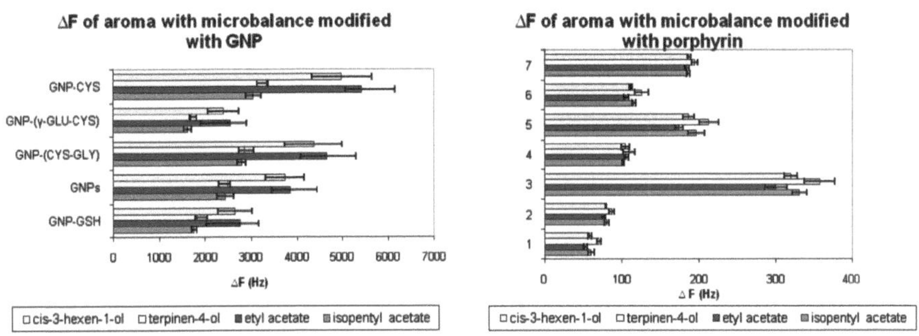

Fig. 8.1 Bar charts of e-nose data using peptide-GNP and porphyrin based sensors

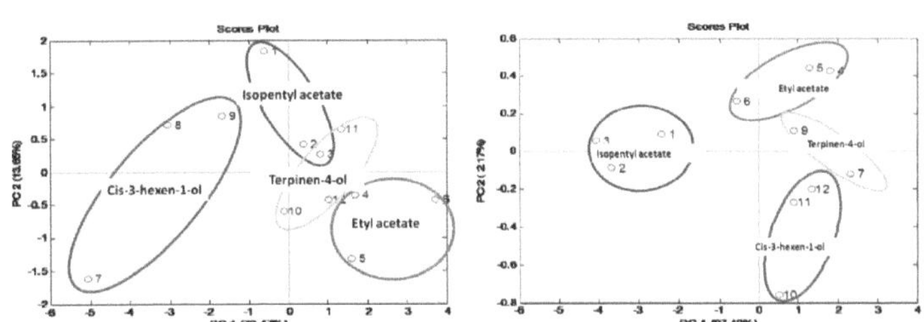

Fig. 8.2 PCA score plots of porphyrin based electronic nose (*left*), and of GNP-peptide electronic nose

4 Conclusions

This preliminary work reports the development of QCM sensors based on gold nanoparticles modified with aminoacids and peptides. The data demonstrate that a simple modification with the tripeptide glutathione, the aminoacid cysteine and the two dipeptides (γ-glutamyl-cysteine) and cysteinylglycine, produce a small set of sensors able to discriminate the typical food aromas tested. The ease of preparation together with the high number of functionalities that are potentially achievable using peptides, makes this approach promising for the application in process control in the food industry.

Acknowledgments The PIRSES-GA-2008-230849 Biomimic project is gratefully aknowledged

References

1. Aryal S, Remant BKC, Dharmaray N, Bhattarai N, Kim CH, Kim HY (2006) Spectroscopic identification of S-Au interaction in cysteine capped gold nanoparticles. Spectrochim Acta A 63:160–163
2. Martinelli E, Mascini M, Santonico M, Pennazza G, Monti D, Paolesse P, D'Amico A, Compagnone D, Di Natale C (2008) Proceeding of IEEE Sensor 26–29 2008, Ottobre Lecce
3. Santonico M, Pittia P, Pennazza G, Martinelli E, Bernabei M, Paolesse R, D'Amico A, Compagnone D, Di Natale C (2008) Study of the aroma of artificially flavoured custards by chemical sensor array fingerprinting. Sensors Actuators B 133:345–351

Part II
Chemical Sensors

Chapter 9
Relative Permittivity of Nanostructured Solid Solutions of Tin and Titanium Oxides

A. Giberti, A. Cervi, and C. Malagù

Impedance Spectroscopy technique was employed in order to characterize nanostructured powders based on solid solutions of SnO_2 and TiO_2, employed in thick-film chemoresistive gas sensors. The measurements were performed with a cylindrical capacitor, manufactured for this specific purpose, with the powders used as dielectric between the plates. Results were interpreted with the support of a theoretical model, which describes the electrical properties of the powders and allows us to estimate their relative permittivity.

1 Introduction

In the field of solid state gas sensors, a great number of semiconductor metal oxides such as ZnO, TiO_2, WO_3 and SnO_2, pure, variously doped or combined in solid solutions show outstanding optical and electronic properties which have attracted attention of the scientific community, in view of their application in the electronic, optoelectronic and photocatalytic fields [1–7]. Among the physical-chemical properties, chemoresistive properties, useful to gas sensing devices [8–14], are very close to the electrical and transport properties, therefore it descends that the dielectric permittivity plays a very important role, because it quantifies the ability of a material to screen the electric fields. The electric field which extends from the

A. Giberti (✉)
Department of Physics, University of Ferrara, Ferrara, Italy

MIST E-R S.C.R.L, Bologna, Italy
e-mail: giberti@fe.infn.it

A. Cervi • C. Malagù
Department of Physics, University of Ferrara, Ferrara, Italy

A. D'Amico et al. (eds.), *Sensors and Microsystems: AISEM 2011 Proceedings*, Lecture Notes in Electrical Engineering 109, DOI 10.1007/978-1-4614-0935-9_9,

surface of the grain of the chemoresistive material into the bulk, is crucial to determine the electrical resistance of the sensitive film [15], therefore a precise knowledge of the permittivity of the material is of fundamental importance. Solid solutions of SnO_2 and TiO_2, synthesized with various tin-titanium proportions, have demonstrated a great potential to detect reducing gases, and do not suffer very much from humidity influence, on the contrary to SnO_2 [16–21]. For this material, a detailed investigation on the relative permittivity has not been performed yet. To this end, Impedance Spectroscopy (IS) technique has gained a great importance during the years as a methodology for the characterization of the materials, and it can be employed to investigate the surface and bulk dielectric response of any kind of solid or liquid material (ionic, semiconducting or insulating).

2 Experimental

Dedicated experiment was performed with a set of powders of SnO_2, TiO_2 and solid solutions of the two oxides.

2.1 *Preparation of Sensing Materials*

The synthesis process for TiO_2, SnO_2 and mixed solid solutions is described in [12]. The $Ti_xSn_{1\text{-}x}O_2$ solid solutions, with $x = 0.3$ and $x = 0.5$, will be hereinafter labeled as ST30 and ST50, respectively.

2.2 *Impedance Spectroscopy Measurements*

IS measurements were performed with the help of a SI 1260 Impedance/Gain-Phase Analyzer by Solartron Instruments. The dielectric response of the powders was analyzed in the range 1 Hz–5 MHz, with the help of a homemade cylindrical capacitor. The powders were employed as dielectric material between the plates of the capacitor (see Fig. 9.1). The theoretical capacitance of the device shown in Fig. 9.1, calculated with the formula for an ideal cylindrical capacitor, is $C_0 = 25\,\text{pF}$. The volume between the plates (2.8 cm^3) can be totally filled with 3 g of powder. All IS measurements were carried out at room temperature (≈25°C), and the data was presented as Cole-Cole plots [22]. The electrical resistance and capacitance of the material were obtained fitting the data with a software, with the help of a suitable model for grain contacts.

3 Results and Discussion

The impedance analyzer was employed to measure the capacitance of the empty cylinder before each material was inserted between the plates. It resulted 31 pF, slightly different from the theoretical value of 25 pF reported above. In Fig. 9.2 the imaginary vs real part of the measured impedance are shown for the materials together with the fits. As can be seen, the fitting circuit of Fig. 9.2 (d) is different from the other three. Indeed, for tin dioxide we used a three-element circuit, whereas for the other materials it was sufficient for an optimal fit to employ just a parallel RQ, where Q is the constant phase element (CPE), in series with a parallel RC. In the case of tin dioxide the corresponding capacitance is C_3, whereas for the other materials it is CPE_2, which is not a pure capacitance due to the non-uniform granular distribution. In fact, the impedance of the CPE is $Z = [(j\omega)^n Y_0]^{-1}$, where ω is 2π times the frequency, j is the imaginary unit and Y_0 is the reactance. In our case the values of n turned out to be about 0.8, therefore the reactive characteristics are dominant and we consider Y_0 as a capacitance. The numerical fit results are summarized in Table 9.1.

We interpreted the results within a model in which every grain possesses the average radius, R, and is considered as a parallel RC circuit: the capacitance ascribed to the material filling the cylindrical capacitor is the result of the series and parallels between these elements. With reference to Fig. 9.1, the number of grain contacts that lie between a and b along the radial direction is $n = (b/2 - a/2)/(2R) - 1$, whereas all the grains that lie on a cylindrical surface of radius $r_i = a/2 + 2R(i + 1)$, where i is an integer between 0 and n, are in parallel.

Consider an imaginary cube of size $2R$ which contain one grain, then the number of grains on the ith surface is $2\pi r_i L/(4R^2)$ and they are all in parallel. The equivalent capacitance of the object is then calculated as:

$$\frac{1}{C_{eq}} = \frac{4R^2}{2\pi L C_{gb}} \sum_{i=0}^{n} \frac{1}{a/2 + 2R(i+1)}, \tag{9.1}$$

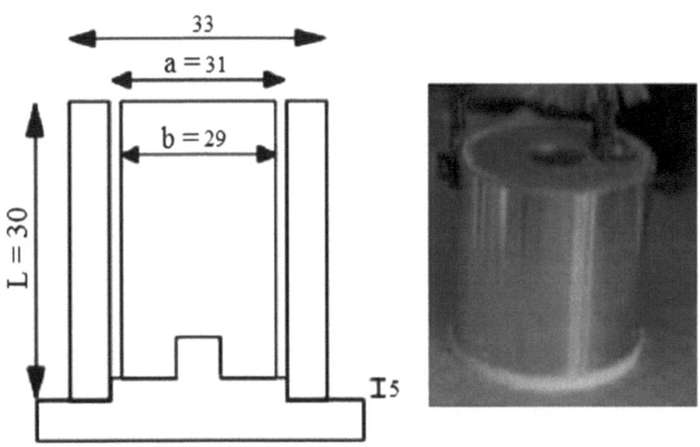

Fig. 9.1 *Left*: sketch of the homemade capacitor, dimensions in mm. *Right*: picture of the device

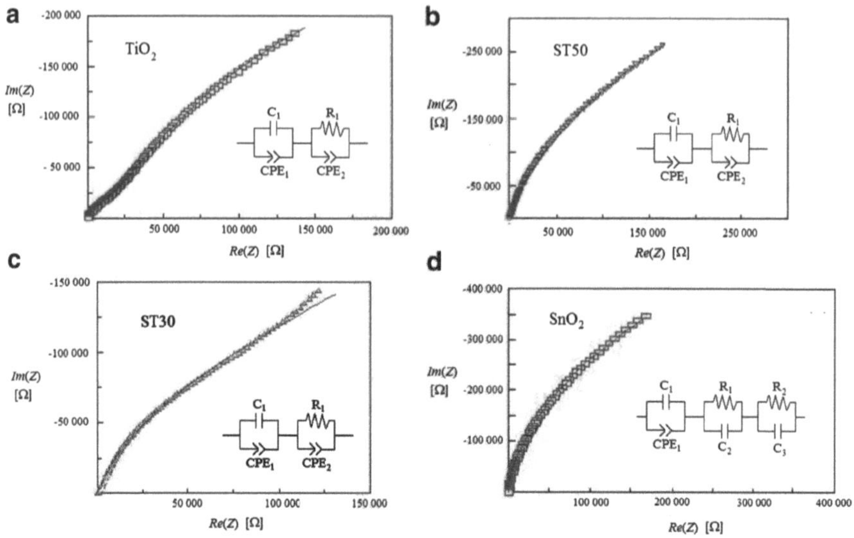

Fig. 9.2 Cole-Cole plots and fit results. The fitting circuit is sketched in each graph

Table 9.1 Numerical results of the fit procedure

	C_1	CPE_1	R_1	CPE_2	C_2	R_2	C_3
SnO_2	370 pF	$Y_0 = 5.0$ nF	120 Ω	–	520 pF	180 kΩ	736 pF
		n = 0.60					
ST30	210 pF	$Y_0 = 100$ nF	2.8 kΩ	$Y_0 = 3.0$ nF	–	–	–
		n = 0.40		n = 0.82			
ST50	253 pF	$Y_0 = 70$ nF	3.1 kΩ	$Y_0 = 4.0$ nF	–	–	–
		n = 0.39		n = 0.78			
TiO_2	150 pF	$Y_0 = 4.5$ μF	660 kΩ	$Y_0 = 2.8$ nF	–	–	–
		n = 0.21		n = 0.85			

Table 9.2 Relative permittivity and depletion length results

	R	Λ	ε_r
SnO_2	15	5	10
ST30	30	10	40
ST50	35	11	52
TiO_2	20	14	75

where C_{gb} is the capacitance of the grain boundary. As described in [22], as far as the grain is far from a fully depleted situation, the capacitance ascribed to the grain boundary is approximated in planar geometry by $C_{gb} = 2\varepsilon_0\varepsilon_r R^2/\Lambda$, where ε_0 is the permittivity of the vacuum, ε_r is the relative permittivity of the material and Λ is the depletion width. After substitution in Eq. 9.1, and using the expression for the Schottky barrier in planar geometry, $V = eN_d\,\Lambda^2/(2\varepsilon_0\varepsilon_r)$, where N_d is the density of donors (we used 4×10^{25} m^{-3} [23]), we obtain a system of two equations in the two variables ε_r and Λ. The results of the calculation are reported in Table 9.2. The grain radius R was calculated with the help of SEM analysis.

4 Conclusions

The present work describes a technique suitable to measure the permittivity of dielectric nanostructured materials. We were able to obtain mixed solutions of SnO_2 and TiO_2 which have shown promising sensing characteristics, as discussed in other papers. The permittivity of these mixed solutions was worked out through a dedicated experiment and the introduction of a specific fitting circuit. The results on well-known pure SnO_2 and TiO_2 are in agreement with the literature.

Acknowledgments The authors wish to thank Silvia Cirillo for the thesis work, from which the present analysis started.

References

1. Rougier A, Portemer F et al (1999) Characterization of pulsed laser deposited WO_3 thin films for electrochromic devices. Appl Surf Sci 153:1–9
2. Chatten R, Chadwick AV, Rougier A, Lindan PJD (2005) The oxygen vacancy in crystal phases of WO_3. J Phys Chem B 109:3146–3156
3. Malagù C, Carotta MC, Comini E, Faglia G, Giberti A, Guidi V, Maffeis TGG, Martinelli G, Sberveglieri G, Wilks SP (2005) Photo-induced unpinning of fermi level in WO_3. Sensors 5:594–603
4. Miyayama M, Takenaka T, Takata M, Shinozaki K (2006) Control of nanoscale morphology of oxide crystals using aqueous solution systems. Key Eng Mat 301:211–214
5. Malagù C, Carotta MC, Morandi S, Gherardi S, Ghiotti G, Giberti A, Martinelli G (2006) Surface barrier modulation and diffuse reflectance spectroscopy of MoO_3-WO_3 thick films. Sensor Actuator B 118:94–97
6. Hayden SC, Allam NK, El-Sayed MA (2010) TiO_2 nanotube/CdS hybrid electrodes: extraordinary enhancement in the inactivation of escherichia coli. J Am Chem Soc 132:14406–14408
7. Snaith HJ, Ducati C (2010) SnO_2-based dye-sensitized hybrid solar cells exhibiting near unity absorbed photon-to-electron conversion efficiency. Nano Lett 10:1259–1265
8. Carotta MC, Benetti M, Ferrari E, Giberti A, Malagù C, Nagliati M, Vendemiati B, Martinelli G (2007) Basic interpretation of thick film gas sensors for atmospheric application. Sensor Actuator B 126:672–677
9. Epifani M, Prades JD, Comini E, Cirera A, Siciliano P, Faglia G, Morante JR (2009) Chemoresistive sensing of light alkanes with SnO_2 nanocrystals: a DFT-based insight. Phys Chem Chem Phys 11:3634–3639
10. Carotta MC, Gherardi S, Malagù C, Nagliati M, Vendemiati B, Martinelli G, Sacerdoti M, Lesci IG (2007) Comparison between titania thick films obtained through sol-gel and hydrothermal synthetic processes. Thin Solid Films 515:8339–8344
11. Roescu R, Dumitriu I, Tomescu A (2004) Simultaneous evaluation of the electrical resistance and work function changes for chemoresistive type sensors. Romanian Rep Phys 56:607–612
12. Carotta MC, Guidi AV, Malagù C, Vendemiati B, Martinelli G (2005) Gas sensors based on semiconductor oxides: basic aspects onto materials and working principles. Mater Res Soc Symp Proc 828:173–184
13. Giberti A, Carotta MC, Guidi V, Malagù C, Martinelli G, Milano L (2009) Influence of ambient temperature on electronic conduction in thick-film gas sensors. Sensor Actuator B 137:111–114

14. Giberti A, Benetti M, Carotta MC, Guidi V, Malagù C, Martinelli G (2008) Heat exchange and temperature calculation in thick-film semiconductor gas sensor systems. Sensor Actuator B 130:277–280
15. Malagù C, Carotta MC, Giberti A, Guidi V, Martinelli G, Ponce MA, Castro MS, Aldao CM (2009) Two mechanisms of conduction in polycrystalline SnO_2. Sensor Actuator B 136:230–234
16. Van Geloven P, Honore M, Roggen J, Leppavuori S, Rantala T (1991) The influence of relative humidity on the response of tin oxide gas sensors to carbon monoxide. Sensor Actuator B 4:185–188
17. Malagù C, Benetti M, Carotta MC, Giberti A, Guidi V, Milano L, Martinelli G (2006) Investigation on the humidity effects on SnO_2-based sensors in CO detection. Mater Res Soc Symp Proc 915:249–255
18. Vlachos DS, Skafidas PD, Avaritsiotis JN (1995) The effect of humidity on tin-oxide thick-film gas sensors in the presence of reducing and combustible gases. Sensor Actuator B 25:491–494
19. Carotta MC, Gherardi S, Guidi V, Malagù C, Martinelli G, Vendemiati B, Sacerdoti M, Ghiotti G, Morandi S, Bismuto A, Maddalena P, Setaro A (2008) (Ti, Sn)O_2 binary solid solutions for gas sensing: spectroscopic, optical and transport properties. Sensor Actuator B 130:38–45
20. Cervi A, Carotta MC, Giberti A, Guidi V, Malagù C, Martinelli G, Puzzovio D (2008) Metal-oxide solid solutions for light alkane sensing. Sensor Actuator B 133:516–520
21. Carotta MC, Cervi A, Giberti A, Guidi V, Malagù C, Martinelli G, Puzzovio D (2009) Ethanol interference in light alkane sensing by metal-oxide solid solutions. Sensor Actuator B 136:405–409
22. Giberti A, Carotta MC, Malagù C, Aldao CM, Castro MS, Ponce MA, Parra R (2011) Permittivity measurements in nanostructured TiO_2 gas sensors. Phys Status Solidi A 208:118–122
23. Malagù C, Guidi V, Carotta MC, Martinelli G (2004) Unpinning of Fermi level in nanocrystalline semiconductors. Appl Phys Lett 84:4158–4160

Chapter 10
NO_2 Sensors with Reduced Power Consumption Based on Mesoporous Indium Oxide

Nicola Donato, Thorsten Wagner, Michael Tiemann, Thomas Waitz, Claus-Dieter Kohl, Mariangela Latino, Giovanni Neri, Donatella Spadaro, and Cesare Malagù

We report on sensing properties of ordered mesoporous nanostructures of In_2O_3 synthesized by nanocasting procedure towards NO_2. The nanostructured material shows improved recovery time and higher response compared to non nano-structured material at low operating temperatures (100–150°C) thus allowing the use for low power NO_2 sensors. These properties may be related to fast oxygen in and out propagation facilitated by an enhanced surface accessibility of the nanostructure.

N. Donato (✉)
Department of Matter Physics and Electronic Engineering
University of Messina, Messina, Italy
e-mail: ndonato@unime.it

T. Wagner • M. Tiemann
Department of Chemistry, University of Paderborn, Paderborn, Germany

T. Waitz
Institute of Inorganic Chemistry, Chemical Didactics, Georg-August-Universität, Göttingen, Germany

C.-D. Kohl
Institute of Applied Physics, Justus-Liebig-University, Giessen, Germany

M. Latino
Department of Chemical Science and Technologies, University of Rome Tor Vergata, Rome, Italy

G. Neri • D. Spadaro
Department of Industrial Chemistry and Materials Engineering, University of Messina, Messina, Italy

C. Malagù
Department of Physics, University of Ferrara, Via Saragat 1/c, 44121 Ferrara, Italy

IDASC - Istituto di Acustica e Sensoristica "O. M. Corbino", Italy

A. D'Amico et al. (eds.), *Sensors and Microsystems: AISEM 2011 Proceedings*, Lecture Notes in Electrical Engineering 109, DOI 10.1007/978-1-4614-0935-9_10,

1 Introduction

The employment of nanostructured sensing materials in the development of chemical sensors has given a new impulse on the improvement of the sensors performance in terms of sensitivity and power consumption. Recently it has been shown that ordered mesoporous metal oxides can be used as sensing layer for resistive sensors [1–3]. They can be synthesized by structure replication (nanocasting) utilising mesoporous silica as a rigid matrix [4,5]. This method has opened up new opportunities to obtain mesoporous materials that were formerly not available. Indium oxide is a suitable sensitive material for the detection of oxidizing gases like O_3 and NO_2 [6,7]. At low temperature (100–150°C) it is nearly insensitive to reducing gases [8], and this increases the selectivity towards oxidizing gases, but response/recovery time of nonstructured materials in this low temperature region prohibits their practical use. Here preliminary results on the development of a resistive NO_2 gas sensor based on ordered mesoporous In_2O_3, synthesized by the nanocasting procedure [9] are reported showing higher responses and improved response/recovery times compared to conventional materials.

2 Experiments

Nanocasting procedure used for synthesis of In_2O_3 is reported in detail elsewhere [9]. Summarizing briefly, the nanocasting procedure comprises two consecutive steps. In the first step an ordered mesoporous silica is synthesized by utilization of supramolecular aggregates of amphiphiles as structure director (soft template). In the second step the pores of the rigid silica matrix are filled with an indium oxide precursor ($In(NO_3)_3$) which is converted by thermal treatment into the oxide after drying. The silica matrix is removed by treatment with sodium hydroxide (NaOH).

The mesoporous In_2O_3 has typical pore sizes and pore wall thicknesses of ca. 5 nm as confirmed by N_2-physisorption measurements (not shown). Figure 10.1 shows an SEM image of a nanostructured In_2O_3 particle confirming the ordered periodic pore structure. This as well as a high degree of crystallinity is also confirmed by XRD.

The material was used for the fabrication of resistive gas sensors, by dispersing the mesoporous In_2O_3 powders in deionized water and then depositing them onto alumina substrates with interdigitated platinum electrodes on the front (electrode distance 200 μm) and a platinum heater on the backside. After drying at room temperature, the sensors were treated for 4 h at 300°C in air before sensing tests.

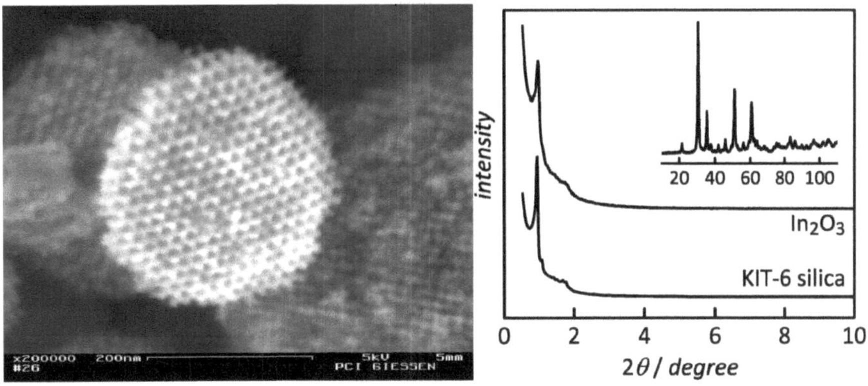

Fig. 10.1 SEM image of the mesoporous In_2O_3 material showing the long-range periodic order of the mesopores. Powder XRD patterns of KIT-6 silica with cubic pore system and its crystalline, mesoporous In_2O_3 replica

3 Results

Preliminary results have shown good performances of the sensors fabricated in the monitoring of low NO_2 concentrations (in the range 0.3–5 ppm). The carried out investigation allowed us to indentify the best working temperature conditions, in a range spanning from 100°C to 300°C. Below 100°C the high electrical resistance of the samples precluded any measurement. As shown in Fig. 10.2, the maximum sensor response to NO_2 concentration of 5 ppm ($R/R_0 > 75$) was obtained at 150°C. These results match well with literature [7].

The dynamical response at this temperature is also shown in Fig. 10.2. Recovery time (time until the sensors signal drops to 10% of its value after end of NO_2 exposure) of the sensor using the mesoporous material is ca. 5 min. This seems much faster than the value obtained in the literature, where the recovery time is larger than 10 min even at 200°C.

In Fig. 10.3 are reported the resistance and the response of the sensor vs. pulses of 10 s in time of NO_2 at different concentrations, at an operating temperature of 150°C.

The high response and the fast recovery times observed appear to be correlated with the nanostructure of the material according to a conductivity model based on atomic oxygen in- and out- propagation. At this point we cannot clarify if diffusion or grain boundary effects can be held responsible for this effect. Similar effects are currently under study for SnO_2 [10] where, however, the oxygen diffusion in the bulk phase is much faster than for In_2O_3. The mesoporous In_2O_3 exhibits oxidizing response to NO_2, which is consistent with the hypothesis of the model that oxygen vacancies are dominant *n*-type defects. However, the electronic transport mechanism in this material is still not understood and FTIR analyses are being carried out together with UV/Vis light tests, to validate the hypothesis.

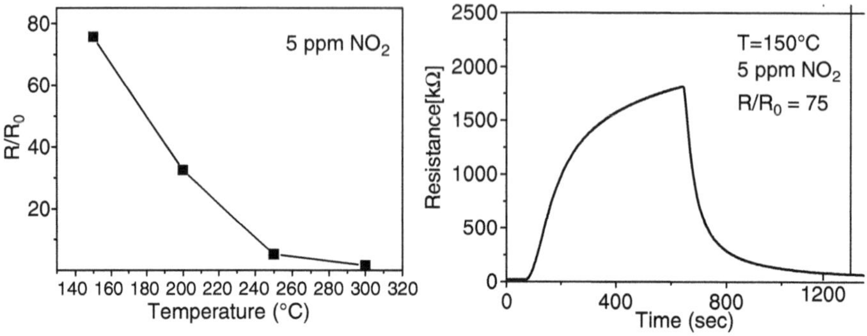

Fig. 10.2 Sensing response of the device to a 5 ppm NO_2 pulse vs. operating temperature. The dynamical response at 150°C is also shown

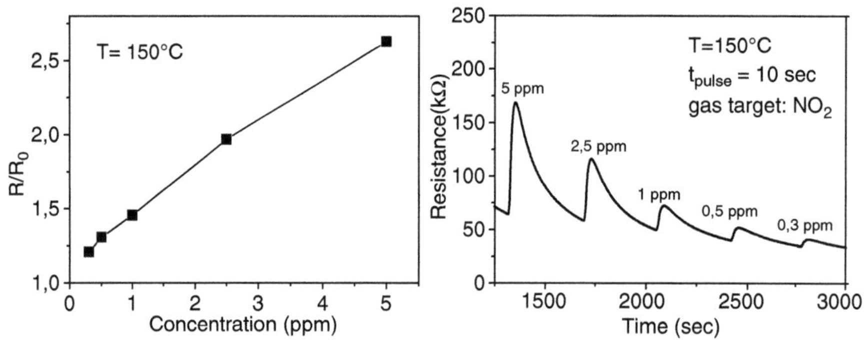

Fig. 10.3 Sensing response and resistance of the device vs. concentration of NO_2 pulses of 10 s

4 Conclusion

The electrical characterization reported in the paper shows that the sensing material is promising for the development of gas sensors requiring reduced power consumption. Further investigations are in progress to test the reliability of the sensor, and to develop strategies aimed to decrease the operating temperature, for example by carrying out electrical characterization under UV light illumination and developing sensors based on MEMs hot plate devices for low power configurations.

References

1. Wagner T, Kohl C-D, Fröba M, Tiemann M (2006) Gas sensing properties of ordered mesoporous SnO_2. Sensors 6:318–323
2. Wagner T, Waitz T, Roggenbuck J, Fröba M, Kohl C-D, Tiemann M (2007) Ordered mesoporous ZnO for gas sensing. Thin Solid Films 515:8360–8363

3. Wagner T, Sauerwald T, Kohl C-D, Waitz T, Weidmann C, Tiemann M (2009) Gas sensor based on ordered mesoporous In_2O_3. Thin Solid Films 517:6170–6175
4. Lu A-H, Schüth F (2006) Nanocasting: a versatile strategy for creating nanostructured porous materials. Adv Mater 18:1793–1805
5. Tiemann M (2008) Repeated templating. Chem Mater 20:961–971
6. Takada T, Suzuki K, Nakane M (1993) Highly sensitive ozone sensor. Sensor Actuator B: Chem 13:404–407
7. Ivanovskaya M, Gurlo A, Bogdanov P (2001) Mechanism of O_3 and NO_2 detection and selectivity of In_2O_3 sensors. Sensor Actuator B: Chem 77:264–267
8. Gurlo A, Barsan N, Ivanovskaya M, Weimar U, Göpel W (1998) In_2O_3 and MoO_3–In_2O_3 thin film semiconductor sensors: interaction with NO_2 and O_3. Sensor Actuator B: Chem 47:92–99
9. Waitz T, Wagner T, Sauerwald T, Kohl C, Tiemann M (2009) Ordered mesoporous In_2O_3: synthesis by structure replication and application as a methane Gas sensor. Adv Funct Mater 19:653–661
10. Ponce MA, Malagù C, Carotta MC, Martinelli G, Aldao CM (2008) Gas indiffusion contribution to impedance in tin oxide thick films. J Appl Phys 104:054907

Chapter 11
Humidity and Temperature Sensors on Flexible Transparency Sheets

G. Scandurra, A. Arena, C. Ciofi, G. Saitta, and G. Neri

1 Introduction

In the last few years there has been an increasing demand of relative humidity sensors to be employed in a variety of applications ranging from process control in manufacturing industries to indoor air quality monitoring in working places and outdoor environmental monitoring. In addition, as the response of most sensing materials towards their specific analyte is affected by humidity, the development of low power consuming sensors capable of operating at low temperature [1, 2] has stimulated a renewed interest toward humidity sensing devices. Flexible RH sensors requiring low power and having light weight, wide working range, and suitably fast recovery/response times, are ideal candidates to be used in tandem with low temperature gas sensors affected by RH interference. Recently inexpensive RH sensors having high flexibility and good sensitivity and stability have been developed using as sensitive materials organic polymers [3], polymers mixtures [4], and polymer/nanoparticles composites [5]. In addition polymer humidity sensors and volatile organic compounds sensors have been integrated on flexible polyimide platforms, equipped with integrated platinum heaters and temperature sensors [6]. In this paper the authors describe a capacitive humidity sensor obtained by depositing zinc-iron-oxide nanopowder embedded in polymethylmethacrylate (PMMA) onto copier grade transparency sheets coated with copper. As far as its response towards humidity is concerned, PMMA is not highly hygroscopic if compared to other organic polymers: it is capable of absorbing a maximum amount of water of

G. Scandurra • A. Arena • C. Ciofi • G. Saitta
Dipartimento di Fisica della Materia e Ingegneria Elettronica,
Università di Messina, Messina, Italy

G. Neri (✉)
Dipartimento di Chimica Industriale e Ingegneria dei Materiali,
Università di Messina, Messina, Italy
e-mail: gneri@unime.it

A. D'Amico et al. (eds.), *Sensors and Microsystems: AISEM 2011 Proceedings*,
Lecture Notes in Electrical Engineering 109, DOI 10.1007/978-1-4614-0935-9_11,

about 2% by weight [7]. Nevertheless, resistive and capacitive relative humidity sensors have been developed using cross linked PMMA [8], and using dispersions based on PMMA [9]. Here we show that the performances of capacitive humidity sensors based on PMMA can be remarkably improved by blending PMMA with nanosized $ZnFe_2O_4$.

2 Experimentals

Copper films were deposited by means of vacuum thermal evaporation, onto copier grade transparency sheets. The coated sheets were patterned (the copper spiral in Fig. 11.1a is an example) through automated application of an etchant ink.

The sensors were developed by applying sensing layers consisting of $ZnFe_2O_4$ nanopowder dispersed in PMMA, onto the copper coated flexible substrates. A number of samples having PMMA as sensing layers were also prepared for comparison. Top electrodes were obtained from conducting dispersions of multi walled carbon nanotubes (MWCNTs) in aqueous solutions of hydroxyl propyl cellulose. The capacitance of a typical Cu/$ZnFe_2O_4$:PMMA/MWCNT sensor (as the one shown in Fig. 11.1b), measured at room temperature, at 55% RH, is found to be in the range of 120 pF. Sensing tests were performed using the set-up sketched in Fig. 11.1c. A Sensirion CMOS temperature and humidity sensor inserted into the measurement chamber was used as reference sensor.

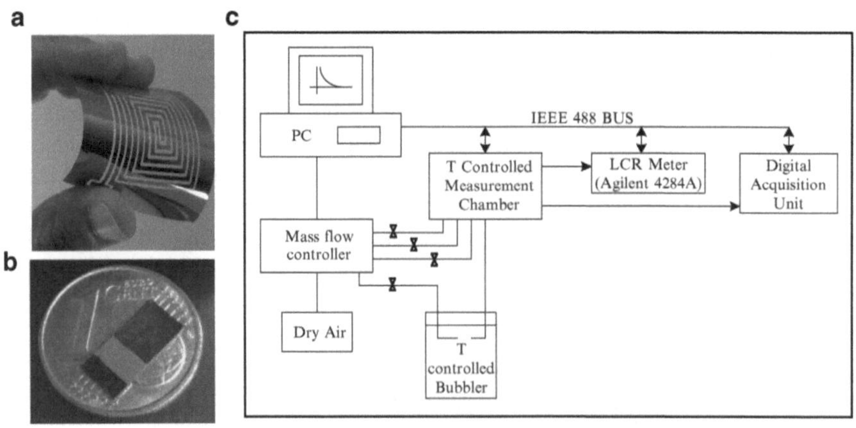

Fig. 11.1 Spiral pattern developed by local chemical etching on a copper coated transparency sheet (**a**); typical example of the developed sensors (**b**); experimental set-up used for sensing tests (**c**)

3 Results and Discussion

Cu/$ZnFe_2O_4$:PMMA/MWCNT and Cu/PMMA/MWCNT sensors having the same design and geometry, have been characterized by means of impedance measurements performed at 23.5°C, at different RH, between 10 Hz and 100 MHz. Figure 11.2a shows the complex impedance of a typical sensor based on $ZnFe_2O_4$:PMMA, measured at 1 kHz, at 23.5°C, as a function of the RH.

According to Fig. 11.2b, a Cu/$ZnFe_2O_4$:PMMA/MWCNT sensor compared to a Cu/PMMA/MWCNT sensor with equal design and geometry, has better performance in terms of dynamic range of capacitance change, in response to RH changes. To demonstrate the reversibility of the response towards humidity of the Cu/ZnFe2O4:PMMA/MWCNT sensor, Fig. 11.3a shows the capacitance measured at 23.5°C, while the RH level inside the measurement chamber is cycled. According to Fig. 11.3a the sensor has response and recovery times lower than 1 min. Figure 11.3b shows the results of static calibration measurements performed at 23.5°C on a typical sensor, fitted to a first order polynomial.

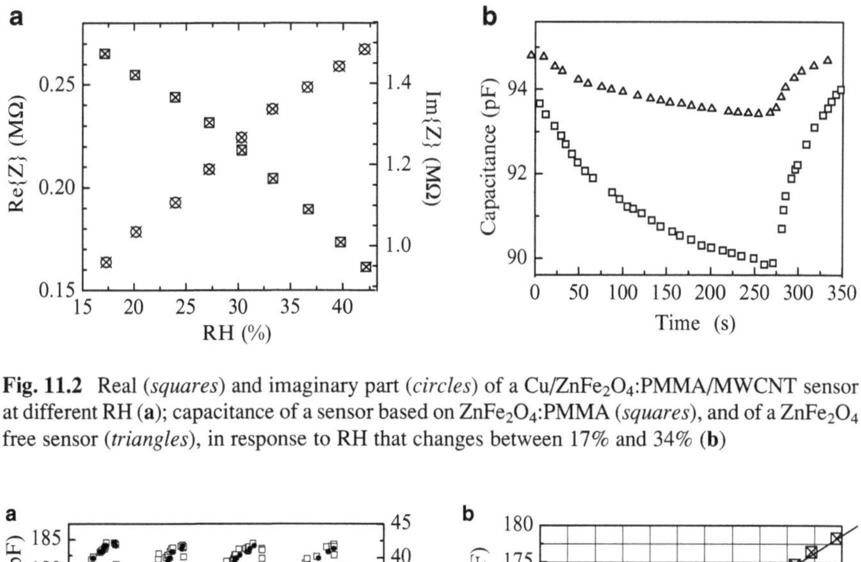

Fig. 11.2 Real (*squares*) and imaginary part (*circles*) of a Cu/$ZnFe_2O_4$:PMMA/MWCNT sensor at different RH (**a**); capacitance of a sensor based on $ZnFe_2O_4$:PMMA (*squares*), and of a $ZnFe_2O_4$ free sensor (*triangles*), in response to RH that changes between 17% and 34% (**b**)

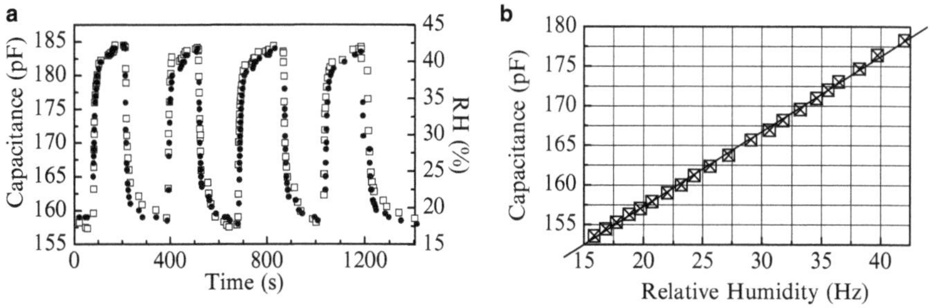

Fig. 11.3 Capacitance of a typical Cu/$ZnFe_2O_4$:PMMA/MWCNT measured as the RH cycles between 17% and 42% (**a**); calibration curve of a typical Cu/$ZnFe_2O_4$:PMMA/MWCNT sensor (**b**)

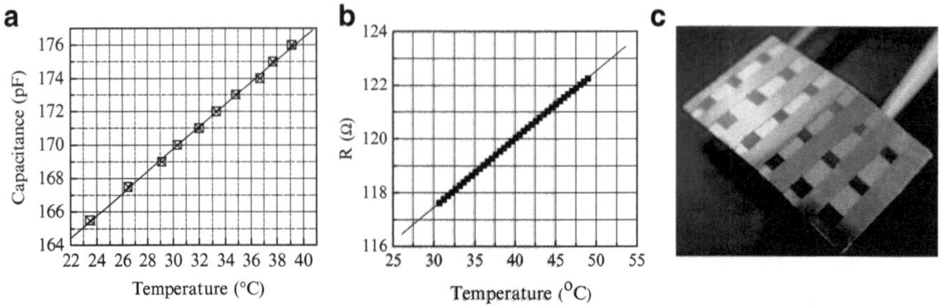

Fig. 11.4 Temperature dependence of the capacitance of a $ZnFe_2O_4$:PMMA sensor, at 17% RH (**a**); temperature dependence of the resistance of a film deposited on flexible sheet (*squares*) and best-fit curve (**b**); copper coated sheet patterned on top and bottom faces using local chemical etching (**c**)

Calibration measurements strictly depend on temperature, as it is found that the capacitance of the Cu/$ZnFe_2O_4$:PMMA/MWCNT sensors measured at constant RH increases as the temperature increases (Fig. 11.4a). This means that the working temperature of the sensors is required to be settled, measured, and controlled in order to provide reliable estimations of RH. Aimed at exploring the possibility to equip the flexible sheets used as substrates with a temperature sensor, four point probes measurements are performed at different temperature on a number of copper films deposited on the sheets. The results show that between room temperature and 70°C, the resistance of the films linearly increases with the temperature (Fig. 11.4b). This finding suggests that once the linear temperature coefficient of resistance is known, copper resistances patterned on the top face, can be used to sense the substrate temperature.

As far as it concerns the temperature of the substrate, it can be adjusted by using copper serpentine heaters, patterned on the bottom face. Automated local application of a $FeCl_3$ based etchant ink is found to be the fastest, easiest and simplest approach to provide double side patterning of double side copper coated sheets (an example is shown in Fig. 11.4c).

4 Conclusions

Inexpensive humidity sensors having simple design and short transient times are developed using transparency sheets as substrates. Aimed at settling and monitoring the substrate temperature, temperature sensors and heaters can be patterned on the bottom and top faces of double side copper coated sheets. Patterning is achieved through a simple approach that minimizes exposure to hazardous chemicals, and avoids time consuming processes including application of photoresist coatings and selective exposure to ultraviolet light.

References

1. Penza M, Rossi R, Alvisi M, Valerini D, Serra E, Paolesse R, Martinelli E, D'Amico A, Di Natale C (2009) Metalloporphyrins-functionalized carbon nanotube networked films for room-temperature VOCs sensing applications Procedia Chem 1:975–978
2. Mattoli V, Mazzolai B, Mondini A, Zampolli S, Dario P (2009) Flexible tag data logger for food logistics. Procedia Chem 1:1215–1218
3. Zampetti E, Pantalei S, Pecora A, Valletta A, Maiolo L, Minotti A, Macagnano A, Fortunato G, Bearzotti A (2009) Design and optimization of an ultra thin flexible capacitive humidity sensor. Sensor Actuator B 143:302–307
4. Oprea A, Bârsan N, Weimar U, Bauersfeld ML, Ebling D, Wöllenstein J (2008) Capacitive humidity sensors on flexible RFID labels. Sensor Actuator B 132:404–410
5. Oikonomou P, Manoli K, Goustouridis D, Raptis I, Sanopoulou M (2009) Polymer/BaTiO3 nanocomposites based chemocapacitive sensors. Microelectr Eng 86:1286–1288
6. Oprea A, Courbat J, Bârsan N, BriandD, de Rooij NF, Weimar U (2009) Humidity and gas sensors integrated on plastic foil for low-power applications. Sensor Actuator B 140: 227–232
7. Rodríguez O, Fornasiero F, Arce A, Radke CJ, Prausnitz JM (2003) Solubilities and diffusivities of water vapor in poly(methylmethacrylate), poly(2-hydroxyethylmethacrylate), poly(N-vinyl-2-pyrrolidone) and poly(acrylonitrile). Polymer 44:6323–6333
8. Matsuguchi M, Yoshida M, Kuroiwa T, Ogura T (2004) Depression of a capacitive-type humidity sensor's drift by introducing a cross-linked structure in the sensing polymer. Sensor Actuator B 102:97–101
9. Su PG, Wang CS (2007) In situ synthesized composite thin films of MWCNTs/PMMA doped with KOH as a resistive humidity sensor. Sensor Actuator B 124:303–308

Chapter 12
Polymer/Metal Oxides Composites on Flexible Commercial Substrates as Capacitive Sensors

N. Donato, D. Aloisio, M. Latino, A. Bonavita, D. Spadaro, and G. Neri

The development of a low level humidity capacitive sensing device working at room temperature, based on a thick layer of iron oxide nanopowders dispersed into a poly(diallyldimethylammoniumchloride) (PDDAC) hydrophilic host matrix deposited on flexible commercial plastic substrates provided with silver electrodes, is reported. The sensor response was tested in a transduction system based on a capacity-frequency conversion of the timing circuit by means of a microcontroller unit. The sensor was investigated in the absolute humidity range from 0% to 1%, showing a good sensitivity and response linearity.

1 Introduction

The development of humidity sensors, working at room temperature, for the monitoring of low humidity levels is an active area of research today due to the great importance of humidity control in many advanced technological applications. Various materials such polymers, polymers mixtures, and polymer/nanoparticles composites, deposited on ceramic or plastic substrates and working as sensing element of humidity sensors, are described in the recent scientific literature [1].

N. Donato (✉) • D. Aloisio
Department of Matter Physics and Electronic Engineering
University of Messina, Messina, Italy
e-mail: ndonato@unime.it

M. Latino
Department of Chemical Science and Technologies
University of Rome Tor Vergata, Rome, Italy

A. Bonavita • D. Spadaro • G. Neri
Department of Industrial Chemistry and Materials Engineering
University of Messina, Messina, Italy

A. D'Amico et al. (eds.), *Sensors and Microsystems: AISEM 2011 Proceedings*, Lecture Notes in Electrical Engineering 109, DOI 10.1007/978-1-4614-0935-9_12,

In this paper is reported our research activity towards the development of low cost gas capacitive sensors and of related characterization system based on a microcontroller and custom electronics.

The devices here presented are based on a thick layer of iron oxide nanopowders dispersed into a poly(diallyldimethylammoniumchloride) (PDDAC) hydrophilic host matrix deposited on flexible commercial plastic substrates provided with silver electrodes. PDDAC is a poly-cationic binder and it has been widely used in combination with metal oxide nanoparticles to produce thick/thin films [2, 3], on the basis of an electrostatic attraction of oppositely charged colloidal particle between the inorganic materials and the polymer [2].

2 Experiments

The sensing material was deposited by drop coating deposition from a water solution of PDDAC and γ-Fe_2O_3 (maghemite) nanopowders. In Fig. 12.1 is reported a SEM micrograph showing the presence of Fe_2O_3 nanoparticles dispersed in the PDDAC host matrix.

The XRD spectrum of the sensing material shown in Fig. 12.2a confirms, by comparison with the diffraction peaks reported in the JCPDS 04–0775 data file, the presence in the PDDAC matrix of the maghemite phase. This is in according with the FT-IR analysis (Fig. 12.2b).

The electrical characterization was performed with an home-made system based on a transduction circuit with NE556 dual monolithic timing devices and a measuring/interfacing one made with an ATMEL AWR Butterfly micro-controller. The transduction system is based on a capacity-frequency conversion of the timing circuit and finally, in a frequency measurement developed by means of the microcontroller unit.

Fig. 12.1 SEM micrograph of the sensing material

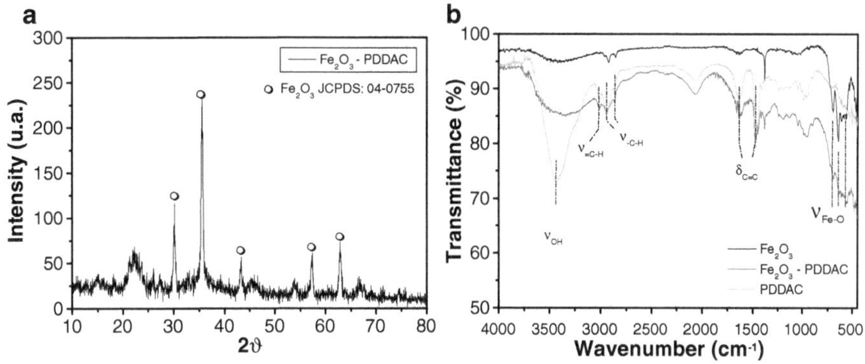

Fig. 12.2 XRD spectrum (**a**) and FT-IR analysis (**b**) of the sensing material

Fig. 12.3 The realized six channel capacitive sensor measurement system

The apparatus shown in Fig. 12.3, is biased by USB Bus and it is interfaced with a personal computer and a Graphical User Interface able to record the frequency and the capacity values of the devices under test.

The system was validated by means of commercial capacitors allowing measurements in a range spanning from 5 pF to 50 nF. The developed system is able to read capacity values of up to six sensors in a single testing procedure. The capacity value of the holder was de-embedded by the capacity ones of the sensors by means of a calibration procedure able to handle the whole sensing array.

3 Results

Humidity sensing test measurements were performed under flux conditions of 50 cc/min, reading the capacity values by varying the water vapor concentration in an absolute humidity range from 0% to 1%.

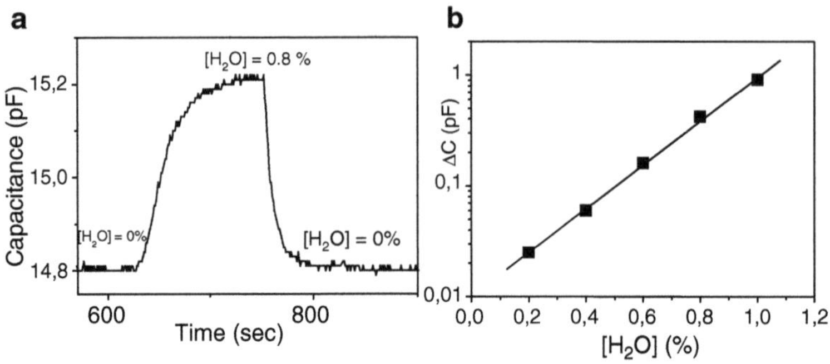

Fig. 12.4 (**a**) Sensor dynamic response to a 0.8% of absolute humidity value. (**b**) Sensor response vs. absolute humidity values

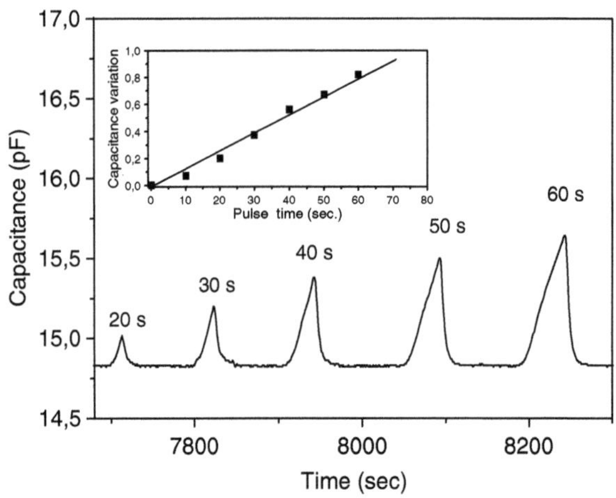

Fig. 12.5 Capacitance response to pulses of water vapour (0.8%) with different time duration. In the *inset*, the correlation between the capacitance variation and pulse time duration is shown

In Fig. 12.4a is reported the sensor dynamic response to a pulse of 0.8% of absolute humidity. It can be noted the well reversible response, with a fast response time and recovery time, in the order of ~ 90 s and ~ 60 s, respectively. In Fig. 12.4b is shown the linear response of the investigated sensor to values of absolute humidity ranging from 0.2% to 0.8%.

Humidity sensing tests were also performed by a pulse method, i.e. maintaining the sensor device in contact with water vapors coming from the bubbler for a lower time than that necessary to reach the complete saturation (Fig. 12.4). In the Fig. 12.5 is reported the capacity values of the sensor tested under pulses of water vapor at a concentration of 0.8% in air with different pulse time, ranging from 20 to 60 s. It can be observed that the capacitance variation increases linearly with pulse time, see inset in the figure, so allowing to find a direct correlation with the humidity concentration.

Providing to establish a suitable calibration, this procedure is particularly effective for practical applications in order to decrease the time between successive pulses, i.e. to increase the number of measurements for unity of time.

4 Conclusion

The development and characterization of a capacitive humidity sensor working at room temperature and of a measurement system able to test sensing arrays composed of up to six sensors, is reported. The sensor was tested in the absolute humidity range from 0% to 1%, showing good sensitivity and linearity. Further activities are in progress to enhance the response of the sensor, by optimizing the sensing materials and the electrodes layout.

References

1. Oprea A, Barsan N, Weimar U, Bauersfeld ML, Ebling D, Wollenstein J (2008) Capacitive humidity sensors on flexible RFID labels. Sensor Actuator B 132:404
2. Arena A, Donato N, Saitta G, Bonavita A, Rizzo G, Neri G (2010) Flexible ethanol sensors on glossy paper substrates operating at room temperature. Sensor Actuator B Chem 145:488–494
3. French RW, Milsoma EV, Moskalenko AV, Gordeev SN, Marken F (2008) Assembly, conductivity, and chemical reactivity of sub-monolayer gold nanoparticle junction arrays. Sensor Actuator B 129:947–952

Chapter 13
Spectroscopy and Electrochemistry of Peptide-Based Self-Assembled Monolayers

M. Caruso, A. Porchetta, E. Gatto, M. Venanzi, M. Crisma, F. Formaggio, and C. Toniolo

Mono and bi-component peptide-based self-assembled monolayers (SAMs) immobilized on a gold surface were studied by electrochemical and spectroscopic techniques. The peptides investigated were exclusively formed by C^{α}-tetrasubstituted amino acids. These residues, due to their peculiar conformational properties, constrain the peptide in a helical conformation, as confirmed by X-ray diffraction structure determinations, and Circular Dichroism and NMR experiments in solution. Both mono-and bi-component peptide SAMs were functionalized with electroactive, fluorescent chromophores strongly absorbing in the UV region. While electrochemical experiments indicated the formation of densely-packed films on the gold surface, fluorescence spectroscopy revealed the occurrence of aromatic-aromatic interactions between the pyrene units functionalizing the peptide chains, obtaining important information on the structural and dynamical properties of the peptide SAMs investigated.

1 Introduction

Hybrid materials obtained by functionalizing metals or semiconductors with biomolecules or bioinspired molecular systems have been recently synthesized, paving the way for the fast-growing field of nanobiotechnology [1]. Among these nanometer scale systems, peptide-based Self-Assembled Monolayers (SAM) have

M. Caruso • A. Porchetta • E. Gatto • M. Venanzi (✉)
Department of Chemical Sciences and Technologies, University of Rome Tor Vergata, Rome, Italy
e-mail: venanzi@uniroma2.it

M. Crisma • F. Formaggio • C. Toniolo
ICB, Padova Unit, CNR, Department of Chemistry, University of Padova, Padova, Italy

A. D'Amico et al. (eds.), *Sensors and Microsystems: AISEM 2011 Proceedings*,
Lecture Notes in Electrical Engineering 109, DOI 10.1007/978-1-4614-0935-9_13,

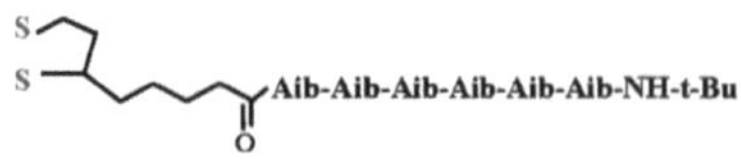

SSA6

SSA6Py

ZAibApi(Pyr)(αMe)ValAib(αMe)Val(αMe)ValAibApi(Boc)NHtBu A8Py

Fig. 13.1 Acronyms and molecular structures of the peptide building-blocks forming the mono- and bi-component SAMs investigated

been investigated as possible substrates for molecular recognition, biocompatible coating for cellular growth, conductive media for Electron Transfer studies over large distances [2].

In this contribution, the properties of peptide-based SAMs linked to gold substrates through Au-S bond were studied by optical spectroscopy (steady-state fluorescence) and electrochemical (Ciclic voltammetry, CV) methods. The peptides investigated were exclusively formed by C^{α}-tetrasubstituted amino acids (Fig. 13.1). These residues, due to their peculiar conformational properties, constrain the peptide in a helical conformation, as confirmed by X-ray diffraction structure determinations, and Circular Dichroism and NMR experiments in solution [3]. An Aib (α-aminoisobutyric acid) homo-hexapeptide was functionalized at the N-terminus with a lipoic group for immobilization to a gold substrate exploiting the strong Au-S affinity ($\approx$40 kcal·mol^{-1}). The peptide was further functionalized with a pyrene chromophore (**SSA6Py**) strongly absorbing in the UV–vis region to enhance the molecular photon capture cross-section of the SAM (*antenna effect*). A peptide with the same backbone, but lacking the pyrene chromophore (**SSA6**), was also synthesized as a reference compound. Furthermore, a photoactive octapeptide (**A8Py**), also formed by C^{α}-tetrasubstituted residues and comprising a pyrene chromophore but lacking the lipoic group, was prepared for obtaining a bi-component peptide SAM formed by inclusion of **A8Py** into the palisade of the **SSA6** SAM linked to the gold surface by Au-S interaction.

2 Results and Discussion

2.1 Cyclic Voltammetry (CV) Experiments

The formation and stability of the **SSA6Py** and **SSA6** SAMs on the gold electrode was checked by CV measurements in the presence of an electrochemical standard redox pair [$K_3Fe(CN)_6$, E°(Fe^{3+}/Fe^{2+} = 0.36 V)] (Fig. 13.2). The deposition of the peptide film partially passivated the gold surface, inhibiting the discharge of the redox pair to the electrode. The decreased activity of the redox pair can be directly related to the package density of the peptide film on the gold surface. Both the **SSA6**

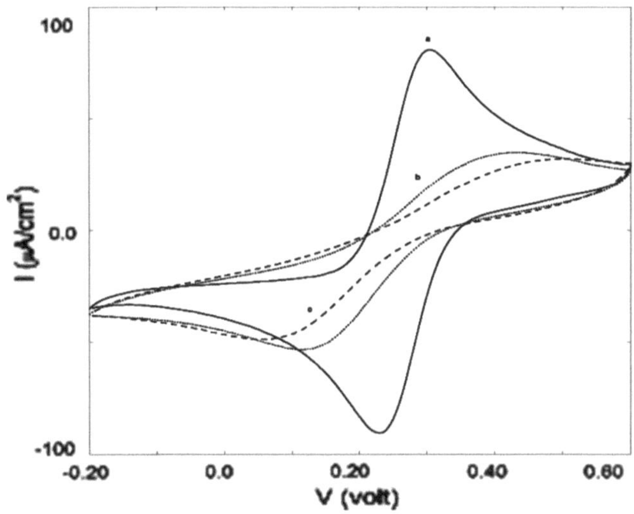

Fig. 13.2 Cyclic voltammetry experiments in a 0.50 mM $K_3[Fe(CN)_6]$. (**a**) bare gold electrode; (**b**) gold electrode modified by SSA6 and (**c**) SSA6Py peptide SAMs

and **SSA6Py** SAMs inhibited the $Fe_3(CN)_6^{3-}$discharge on the gold electrode, although a residual capacitive current, most likely ascribable to diffusion of the buffer electrolyte (KCl), was still measured for the modified electrode. Interestingly, for the **SSA6/A8Py** bi-component SAM the discharge of the redox pair was found to be almost completely depleted indicating the formation of a densely-packed SAM.

CV experiments also showed that the Pyrene group in the **SSA6Py** SAM gave rise to irreversible oxidation at 0.95–1.0 V. After that, a new peak at 0.2–0.4 V can be observed, ascribable to the discharge of diol/diketone species, stable byproducts of Pyrene oxidation. This peak could be observed after repeated scans, signaling the integrity of the peptide SAM on the gold surface at these applied potentials. Disruption of the Au–S linkages was only observed at negative applied potentials (≤ -0.8 V).

2.2 *Steady-State Fluorescence Experiments*

Peptides functionalized with Pyrene chromophores allowed for easy determination of the onset of interchain interactions, due to the characteristic emission properties of pyrene groups. While the monomer emission is characterized by well-resolved vibrational transitions, pyrene-pyrene excited state interaction gives rise to a broad and intense red-shifted emission associated to the formation of dimeric excited-state complexes (*excimers*). As can be observed in Fig. 13.3, the emission spectrum of the **SSA6Py** SAM, linked to a 5 nm gold film supported on quartz, showed a

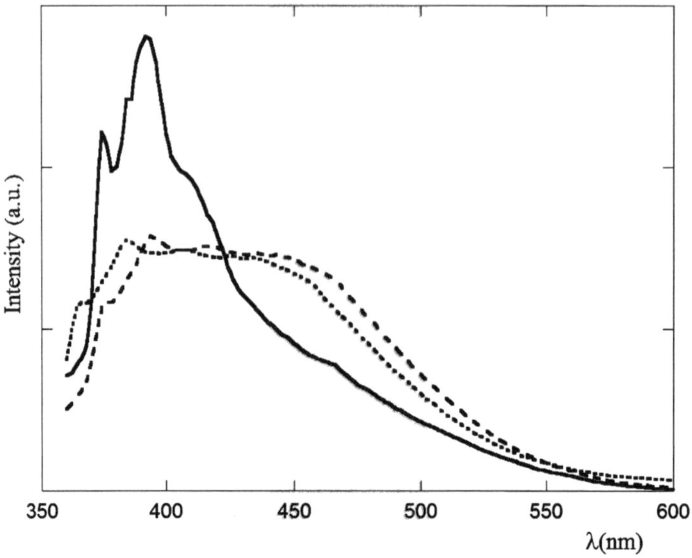

Fig. 13.3 Emission spectra of mono-and bi-component peptide SAMs: SSA6Py (*dashed line*); SSA6Py:SSA6 (1:10) (*continuous line*); A8Py:SSA6 (1:1) (*dotted line*)

broad red-shifted fluorescence, indicating the formation of excimer species. On the contrary, the fluorescence spectrum of the bi-component SAM obtained by a (1:10) **SSA6Py/SSA6** millimolar deposition solution, also reported in Fig. 13.3, was characterized by the typical emission of pyrene monomer species.

Interestingly, the fluorescence spectrum of the bi-component SAM formed by a (1:1) **A8Py** and **SSA6** millimolar deposition solution revealed an excimer-like emission, as also shown in Fig. 13.3. This finding confirmed the inclusion of **A8Py** in the **SSA6** palisade, linked to the gold surface by the strong Au-S electrostatic interaction. The densely-packed nature of this bi-component SAM, stabilized by favorable dipole-dipole interaction between the **A8Py** and **SSA6** peptide chains, was confirmed by the CV experiments.

The observation of excimer emission strongly suggests the formation of **A8Py** segregated domains (*rafts*) within the **SSA6** SAM. This effect is most likely ascribable to the dynamic nature of the processes leading to the formation of self-assembled monolayers. The relatively free diffusion of **A8Py**, lacking the lipoic group, allowed for the slow organization of **A8Py** domains during the SAM deposition (18 h). This conclusion was strengthened by the absence of excimer-like emission in bi-component SAMS formed by (1:1) **SSA6/SSA6Py** millimolar deposition solution, both strongly linked to the gold surface through Au-S interaction (data not shown).

3 Conclusions

The packing density and stability of mono-and bi-component peptide-based self-assembled monolayers were characterized by electrochemical and spectroscopic measurements. Cyclic Voltammetry experiments showed that all the peptide SAMs are densely-packed despite the shortness of the oligopeptides used as building blocks in the self-assembly process. This is probably due to the conformationally-constrained character of peptides rich in C^{α}-tetrasubstituted amino acids. Fluorescence experiments revealed that aromatic-aromatic interactions contribute to the stabilization of the peptide film on the electrode surface, forming separated domains whenever possible. The characteristic emission of excited state complexes (excimer) was exploited for monitoring the onset of interchain interactions between the peptide building blocks.

Acknowledgments The financial support (PRIN 2008, Project 20088NTBKR) of the Italian Ministry for University and Research (MIUR) is acknowledged.

References

1. Niemeyer CM, Mirkin CA (eds) (2004) Nanobiotechnology. Wiley-VCH, Weinheim
2. Bianco A, Venanzi M, Aleman C (2011) Peptide-based materials: from nanostructures to applications. J Pept Sci 17:73–74
3. Gatto E, Caruso M, Porchetta A, Toniolo C, Formaggio F, Crisma M, Venanzi M (2011) Photocurrent generation through peptide-based self-assembled monolayers on a gold surface: antenna and junction effects. J Pept Sci 17:124–131

Chapter 14
Organic Vapor Detection by QCM Sensors Using CNT-Composite Films

M. Alvisi, P. Aversa, G. Cassano, E. Serra, M.A. Tagliente, M. Schioppa, R. Rossi, D. Suriano, E. Piscopiello, and M. Penza

A Quartz Crystal Microbalance (QCM) gas sensor coated with carbon nanotubes (CNTs) layered films as chemically interactive nanomaterial is described. A QCM resonator integrated on AT-cut quartz substrate has been functionally characterized as oscillator at the resonant frequency of 10 MHz. The CNTs have been grown by chemical vapor deposition (CVD) system onto alumina substrates, coated with 2.5 nm thick Fe catalyst, at a temperature of 750°C in H_2/C_2H_2 gaseous ambient as active materials for gas sensors. CNTs multilayers, with and without buffer layer of cadmium arachidate (CdA), have been prepared by the Langmuir-Blodgett (LB) technique to coat at the double-side the QCM sensors for organic vapor detection, at room temperature. It was demonstrated that the highest mass sensitivity has been achieved for CNTs multilayer onto CdA buffer material due to the greatest gas adsorbed mass. The sensing properties of the CNTs-sensors at enhanced mass sensitivity have been investigated for different vapors of ethanol, methanol, acetone, m-xylene, toluene and ethylacetate in a wide range of concentration from 10 to 800 ppm. The CNTs-based QCM-sensors exhibit high sensitivity (e.g., 5.55 Hz/ppm to m-xylene of the CNTs-multilayer) at room temperature, fast response, linearity, reversibility, repeatability, low drift of the baseline frequency, potential sub-ppm range detection limit.

1 Introduction

Quartz Crystal Microbalances (QCMs) resonators have been widely used as high-performance transducers and promising sensor platforms for chemical detection of targeted analytes in air [1–3] and/or water [1, 4] phase. The QCM sensors are very

M. Alvisi • P. Aversa • G. Cassano • E. Serra • M.A. Tagliente
• M. Schioppa • R. Rossi • D. Suriano • E. Piscopiello • M. Penza (✉)
ENEA, Brindisi Technical Unit for Technologies of Materials, Brindisi, Italy
e-mail: michele.penza@enea.it

A. D'Amico et al. (eds.), *Sensors and Microsystems: AISEM 2011 Proceedings*, Lecture Notes in Electrical Engineering 109, DOI 10.1007/978-1-4614-0935-9_14,

interesting for practical applications owing to their high sensitivity and ease of measurement. QCM measurements rely on the Sauerbrey equation, which relates the change in the resonant frequency of a quartz crystal to the change in mass due to gas adsorption on the crystal. The frequency shift of a QCM sensor is strongly depending on squared frequency, typically ranging from 5 to 30 MHz, and surface density ($\Delta m/A$) of the sensitive film. In fact, the mass sensitivity of a QCM resonating at 10 MHz is typically in the range of 0.1 - 1.0 Hz/(ng/cm^2), which is very high to detect very-low concentrations of gases up to sub-ppm level.

Therefore, QCM sensitivity can be improved by mass amplification using very adsorbent materials which increase the adsorbed mass. In this context, nanomaterials with high surface area and high chemical reactivity are very useful for fabricating high-sensitivity chemical sensors. Gas sensors based on carbon nanotubes (CNTs) have been studied both in the form of nanocomposites [5–9] and networked films [10, 11] for high-sensitive VOCs detection SAW and QCM applications, at room temperature.

Here, *CNTs-in-CdA* composite and *CNTs-on-CdA* layered films have been prepared by Langmuir-Blodgett (LB) and casting technique onto double-side AT-cut quartz 10 MHz QCM equipped by Al electrodes for Volatile Organic Compounds (VOCs) detection at ppm level in the range of related Threshold Limit Value (TLV) of ethylacetate, m-xylene, toluene, acetone, alcohols, at room temperature. A buffer layer of Cadmium Arachidate (CdA) has been used to promote the adhesion of the CNTs multilayer onto QCM surface and as host-matrix in the composite with weight-controlled filler of CNTs. These carbon-based nanomaterials were grown by CVD technology at ENEA laboratories.

2 Experimental Details

CNTs films were grown by CVD technology. The CNTs films were deposited onto large-size cost-effective alumina (40 mm width x 40 mm length x 0.6 mm thickness), coated with growth-catalyst of iron (Fe) nanoclusters with a nominal thickness of 2.5 nm and sputtered at 10^{-2} Torr. The Fe-catalysed alumina substrates were heated to 750°C by a rate of 10°C/min in H_2 atmosphere upon flow of 100 sccm at a total pressure of 1.5 Torr. In the gas-plasma, the flow rate ratio between C_2H_2 and H_2 was kept constant at 20/80 sccm, respectively. The CNTs deposition was performed at a constant pressure and temperature of 5 Torr and 750°C, respectively for 30 minutes by depositing a vertically-aligned CNTs film with thickness of 10–12 μm.

After CNTs growth, the nanomaterial was mechanically removed from substrate to prepare a solution as precursors for the LB film deposition. A dispersion of the CNTs in a DMF solvent has been prepared to promote their de-bundling before LB film deposition. In addition, a buffer layer of cadmium arachidate (CdA) has been used to promote the adhesion of the CNTs multilayer onto QCM surface.

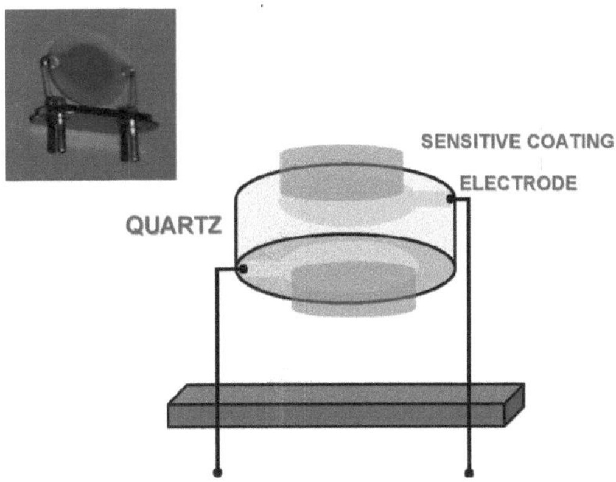

Fig. 14.1 Scheme of the QCM sensor measured and image of a typical QCM

The scheme of the QCM sensor is shown in Fig. 14.1. A piezoelectric sensor based on unloaded AT-cut quartz oscillating at a frequency of 10 MHz has been used as transducer equipped by Al electrodes with an active area of 12.56 mm^2.

Different types of QCM-sensors have been fabricated by depositing LB layered films in the format CNTs-on-CdA multilayer and CNTs-in-CdA composite. The process parameters for LB film deposition are reported elsewhere [5, 6, 8].

The so-fabricated QCM sensors have been located in a test cell (1,500 mL volume) for gas exposure measurements. The cell case is able to host up to twelve piezoelectric sensors. Dry air was used as reference gas and diluting gas to air-conditioning the sensors. The gas flow rate was controlled by mass flowmeters. The total flow rate per exposure was kept constant at 1,500 mL/min. The gas sensing experiments have been performed by measuring the resonant frequency of the three QCM sensors upon controlled ambient of individual volatile organic compounds (VOCs) of ethanol, methanol, acetone, m-xylene, toluene and ethylacetate in the range of 30–100 ppm, 50–180 ppm, 240–720 ppm, 12–40 ppm, 25–100 ppm and 100–400 ppm, respectively, at sensor temperature of 20°C. The frequency output of the QCM-sensors has been measured by a frequency counter (Agilent, 53132A) with a multiplexed read-out by a switch unit (Agilent, 34970A) driving two 50 Ω, 4 x 1 rf multiplexers (Agilent, 34905A). A J-type thermocouple was used to control the temperature in the sensor cell and its voltage output was measured by a multimeter (Agilent, 34401A). Data were collected and stored for further analysis in a PC interfaced with a GPIB card in the VEE-software ambient (Agilent).

3 Results and discussion

The morphology of the CNTs-in-CdA composite, where CNTs filler are treated with DMF solvent, has been characterized by FE-SEM, as reported in the Fig. 14.2. A substrate of silicon has been used to deposit DMF-treated CNTs layers for

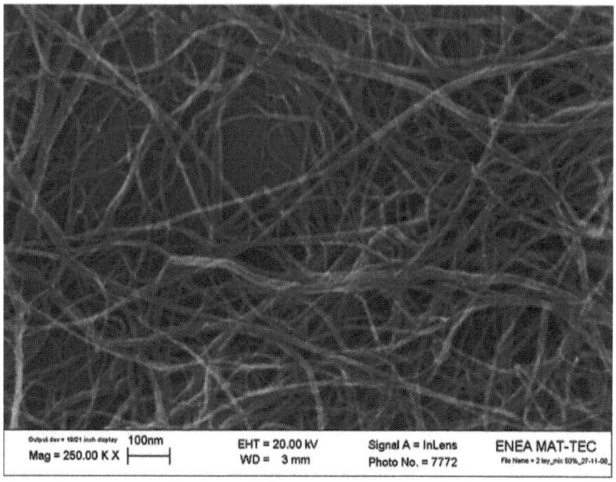

Fig. 14.2 FE-SEM image of the CNTs treated with DMF and prepared on Si substrate by LB deposition of a coating of ten-layer CNTs-in-CdA 50% composite

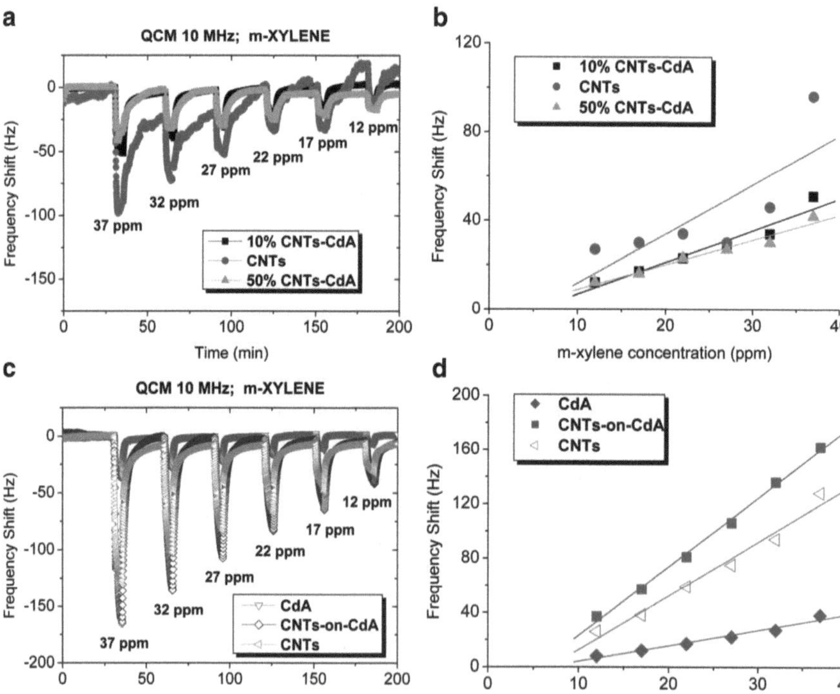

Fig. 14.3 Sensing characteristics of time response towards 5-min six pulses of m-xylene and related calibration curves for the QCM sensors coated by [(**a**), (**b**)] 10% and 50% CNTs-in-CdA Composite, or CNTs layer, and [(**c**), (**d**)] CNTs-on-CdA Multilayer, or CdA buffer, or CNTs layers

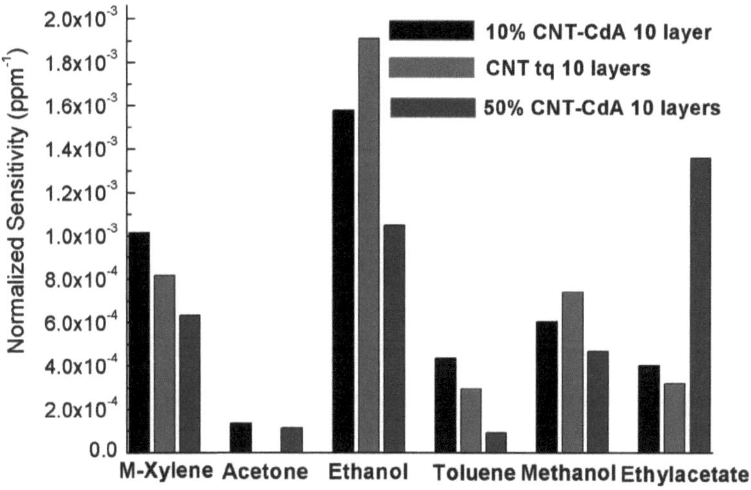

Fig. 14.4 Room-temperature pattern sensitivities of three 10 MHz QCM-sensors coated by 10% and 50% CNTs-in-CdA, or CNTs layer, towards six different VOCs of m-xylene, acetone, ethanol, toluene, methanol and ethylacetate. The range of the tested gas concentration of m-xylene, acetone, ethanol, toluene, methanol and ethylacetate is 12–40 ppm, 240–720 ppm, 30–100 ppm, 25–100 ppm, 50–180 ppm, and 100–400 ppm, respectively. The gas sensitivity has been normalized to the frequency shift due to the mass loading of each sensitive coating

electron microscopy observations. A dense network of bundles of tubes consisting of multi-walled carbon nanostructures appears with a maximum length of 5 μm and single-tube diameter varying in the range of 5–35 nm.

Figure 14.3 shows the typical time responses in terms of frequency shift for three QCM sensors based on LB films of CNTs-in-CdA composite and CNTs-on-CdA multilayer, exposed towards 5-min six pulses of m-xylene, at room-temperature. The resonant frequency of all CNTs-sensors decreases upon a single exposure of the m-xylene caused by the mass loading of the molecules adsorption. All QCM-sensors demonstrate reversibility of the response upon switching of the target analyte concentration into dry air. In addition, a total recovery of the baseline frequency for all QCM-sensors has been measured. The highest mass sensitivity has been achieved for multilayer CNTs-on-CdA due to the greatest gas adsorbed mass. A linearity in the calibration curves for all QCM-sensors was measured, as reported in Fig. 14.3b, d.

The gas sensitivity, expressed by the slope of the linear curves, has been plotted towards six tested different analytes (m-xylene, acetone, ethanol, toluene, methanol and ethylacetate) for all studied composite coatings, as reported in the Fig. 14.4. The gas sensitivity has been normalized to the frequency shift due to the mass loading of each sensitive coating. These results demonstrate that CNTs-in-CdA composite exhibits the maximum sensitivity measured for four VOCs of m-xylene, acetone, toluene and ethylacetate due to its highest capability of gas adsorption. In the contrast, CNTs multilayer exhibits highest sensitivity for two VOCs of ethanol and methanol. A tailored sensitivity towards aromatics (m-xylene, toluene), esters (ethylacetate), ketones (acetone) and alcohols (ethanol, methanol) has been detected for any considered sensitive coating.

The gas adsorption can be considered as the contribution of the main sensing mechanism of mass loading, as measured by the linearity in the gas response in terms of frequency shift as a function of gas concentration.

4 Conclusions

CNTs-composite and CNTs-multilayer for 10 MHz QCM sensors have been prepared by Langmuir-Blodgett technique for enhanced mass sensitivity to detect VOCs of ethanol, methanol, ethylacetate, acetone, toluene and m-xylene up to ppm level of concentration, at room temperature. Chemical sub-ppm level of the VOCs is potentially measurable by tested QCM-sensors based on CNTs-multilayer. The gas sensing properties have been investigated comparing the gas response of the buffered and un-buffered CNTs-based multilayers and CNTs-composites to maximize sensitivity of the CNT sensors. The results demonstrate that the CNTs-on-CdA multilayer exhibits the highest gas sensitivity towards tested VOCs, at room temperature. This is caused by the enhanced mass loading of the adsorbed gas molecules.

The QCM-sensors show high sensitivity, linearity, fast response, potential sub-ppm detection level, good reproducibility and reversibility. The studied QCM sensors with enhanced gas sensitivity and broader selectivity are useful for chemical detection at room temperature and low-power consumption and for environmental air monitoring applications.

References

1. Voinova MV (2009) J Sensors. art. ID 943125
2. Ballantine DS, White RM, Martin SJ, Ricco AJ, Zellers ET, Frye GC, Wohltjen H (1997) Acoustic wave sensors. Academic, San Diego
3. Serban B, Sarin Kumar AK, Costea S, Mihaila M, Buiu O, Brezeanu M, Varachiu N, Cobianu C (2009) Polymer-amino carbon nanotube nanocomposites for surface acoustic wave Co_2 detection. Rom J Inf Sci Technol 12(3):376–384
4. Rabe J, Buttgenbach S, Schroder J, Hauptmann P (2003) Monolithic miniaturized quartz microbalance array and its application to chemical sensor systems for liquids. IEEE Sensors J 3(4):361–368
5. Penza M, Tagliente MA, Aversa P, Cassano G, Capodieci L (2006) Single-walled carbon nanotubes nanocomposite microacoustic organic vapor sensors. Mater Sci Eng C 26:1165–1170
6. Penza M, Tagliente MA, Aversa P, Cassano G (2005) Organic vapor detection using carbon nanotube composites microacoustic sensors. Chem Phys Lett 409:349–354

7. Penza M, Tagliente MA, Aversa P, Re M, Cassano G (2007) The effect of purification of single-walled carbon nanotube bundles on the alcohol sensitivity of nanocomposite Langmuir–Blodgett films for SAW sensing applications. Nanotechnol 18:185502
8. Penza M, Aversa P, Cassano G, Wlodarski W, Kalantar-zadeh K (2007) Layered SAW gas sensor with single-walled carbon nanotube-based nanocomposite coating. Sensor Actuat B 127:168–178
9. Chen H-W, Wu R-J, Chan K-H, Sun Y-L, Su P-G (2005) The application of CNT/Nafion composite material to low humidity sensing measurement. Sensor Actuat B 104:80–84
10. Consales M, Campopiano S, Cutolo A, Penza M, Aversa P, Cassano G, Giordano M, Cusano A (2006) Sensing properties of buffered and not buffered carbon nanotubes by fibre optic and acoustic sensors. Meas Sci Technol 17:1220–1228
11. Su P-G, Tsai J-F (2009) Low-humidity sensing properties of carbon nanotubes measured by a quartz crystal microbalance. Sensor Actuat B 135:506–511

Chapter 15
A Portable Sensor System for Air Pollution Monitoring and Malodours Olfactometric Control

D. Suriano, R. Rossi, M. Alvisi, G. Cassano, V. Pfister, M. Penza, L. Trizio, M. Brattoli, M. Amodio, and G. De Gennaro

A portable sensor-system based on solid-state gas sensors has been designed and implemented as proof-of-concept for environmental air-monitoring applications and malodours olfactometric control. Commercial gas sensors (metal-oxides, n-type) and nanotechnology sensors (carbon nanotubes, p-type) are arranged in a configuration of array for multisensing and multiparameter devices. Wireless sensors at low-cost are integrated to implement a portable and mobile node, that can be used as early-detection system in a distributed sensor network. Real-time and continuous monitoring of hazardous air-contaminants (e.g., NO_2, CO, SO_2, BTEX, etc.) has been performed by in-field measurements. Moreover, monitoring of landfill gas generated by fermentation of wastes in a municipal site has been carried out by the portable sensor-system. Also, it was demonstrated that the sensor-system is able to assess the malodours emitted from a municipal waste site and remarkably compared to the olfactometry method based on a trained test panel.

1 Introduction

A strong demand of cost-effective and high performance gas sensors involves several technological sectors such as air quality control, chemical security and safety, energy applications, environmental monitoring, including malodours detection and olfactometric control.

D. Suriano • R. Rossi • M. Alvisi • G. Cassano • V. Pfister • M. Penza (✉)
ENEA, Brindisi Technical Unit for Technologies of Materials, Brindisi, Italy
e-mail: michele.penza@enea.it

L. Trizio • M. Brattoli • M. Amodio • G. De Gennaro
Department of Chemistry, University of Bari, Lenviros srl, spin-off from University of Bari, Bari, Italy

A. D'Amico et al. (eds.), *Sensors and Microsystems: AISEM 2011 Proceedings*, Lecture Notes in Electrical Engineering 109, DOI 10.1007/978-1-4614-0935-9_15,

Many air pollution systems in urban and rural areas that utilize smart sensor networks and wireless systems have been reported in recent literature [1–4].

Analysis system based on electronic olfaction is considered a powerful technology for portable and handled devices. Various *Electronic Noses* [5–12] have been developed for air quality control and malodours detection.

A portable sensor-system based on solid-state gas sensors has been designed and implemented as proof-of-concept for environmental air-monitoring applications and malodours olfactometric control. Commercial gas sensors (metal oxides, n-type) and nanotechnology sensors (carbon nanotubes, p-type) by ENEA patent have been arranged in a configuration of array for multisensing and multiparameter devices. Wireless sensors at low-cost are integrated to implement a portable and mobile node, that can be used as early-detection system in a distributed sensor network. Real-time and continuous monitoring of hazardous air-contaminants (e.g., NO_2, CO, SO_2, BTEX, etc.) has been performed by in-field measurements. Additionally, monitoring of landfill gas in a municipal waste site has been experimented as well to control methane and non-methanic hydrocarbons (NMHC) generated by fermentation of solid urban wastes. Finally, the portable sensor-system has been used to control malodours emitted from a municipal waste site to assess the odourant impact for real-time and in-situ measurements.

In this study, preliminary results of the experimental campaigns of the portable sensor-system are reported for air quality control, landfill gas monitoring and malodours control.

2 Experimental Details

Validation of the sensing performance of the portable system has been realized by means of the chemical analyzers of the Italian environmental air monitoring agency (ARPA-PUGLIA). A medium-term experimental campaign has been performed and some preliminary results are presented to address the comparison between sensor-system and chemical analyzers regulated by EU standards. The results demonstrate that the sensor-system is a complementary valid tool to realize a low-cost sensor node in a distributed network for environmental-air monitoring applications. Figure 15.1 shows the portable sensor-system developed at laboratories ENEA Brindisi for air quality control and used in a real scenario. The array based on four gas sensors is shown in Fig. 15.2.

3 Results and Discussion

Figure 15.3 shows the typical time responses in terms of sensor voltage of the p-type CNT sensor (ENEA patent) and a commercial n-type MOX sensor (TGS 2106, Figaro) towards NO_2 gas measured in ambient air by a chemiluminescence

Fig. 15.1 Portable sensor-system developed at laboratories ENEA Brindisi and used in real scenario for air quality control, landfill gas monitoring, and malodours control

Fig. 15.2 Sensor array used in the sensor-system for environmental air monitoring. The sensors used in the array are TGS 2106, TGS 2600, TGS 822 (Figaro), and CNT:Pt8nm (ENEA patent)

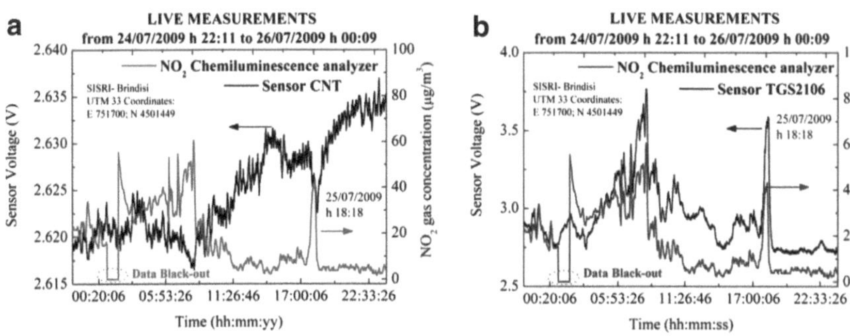

Fig. 15.3 (**a**) Comparison of tracks of an innovative gas sensor (CNTs) and a NO_2 chemical analyzer at an air quality monitoring site. (**b**) Comparison of tracks of a commercial gas sensor (TGS 2106) and a NO_2 chemical analyzer at an air quality monitoring site (ARPA-PUGLIA)

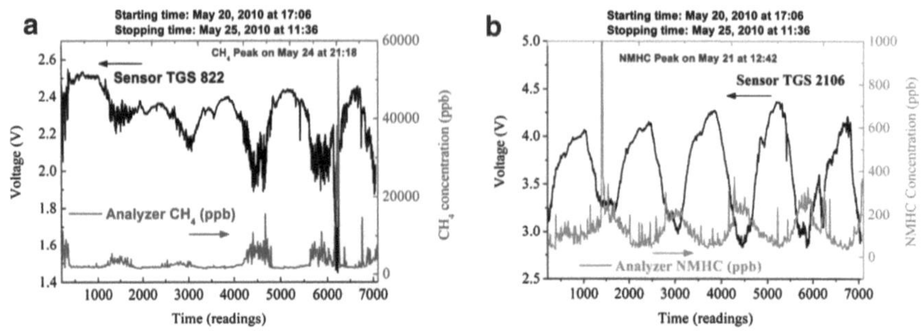

Fig. 15.4 (**a**) Comparison of tracks of a gas sensor (TGS 822) and a chemical analyzer at a waste site for monitoring methane (CH_4) in landfill gas. (**b**) Comparison of tracks of a gas sensor (TGS 2106) and a chemical analyzer at a waste site for monitoring non-methanic hydrocarbons (NMHC) in a landfill gas

analyzer in an air-quality site (Arpa-Puglia). These results demonstrate that both gas sensors are able to follow the real-time NO_2 gas concentration in ambient air with the sensor TGS 2106 more sensitive than un-modified CNT sensor to measure the recorded NO_2 peak of about 40 $\mu g/m^3$ on 25 July 2009 at 18:18 (local time). This value is less than attention level and alarm level, as regulated by Italian regulations for NO_2 of 200 and 400 $\mu g/m^3$, respectively.

Real-time and continuous monitoring of hazardous air-contaminants, greenhouse gases, and landfill gas (LFG), mainly constituted by CO_2 and CH_4 has been performed in field measurements at a municipal waste site. Validation of the sensing performance of the portable system has been realized by means of a chemical analyzer. A short-term experimental campaign has been performed and the results address a performance comparison between portable sensor-system and chemical analyzer regulated by EU standards.

Figure 15.4 indicates a comparison of the response of two commercial n-type gas sensors (TGS 822 and TGS 2106, Figaro) and a chemical analyzer for detection of methane (CH_4) and non-methanic hydrocarbons (NMHC) in a real scenario of a municipal waste site to assess the generated landfill gas. These results demonstrate that TGS 822 sensor is able to monitor CH_4 gas, as shown by the measured peak of about 60,000 ppb (60 ppm), while TGS 2106 sensor is able to monitor the NMHC component in the landfill gas.

Finally, these preliminary results prove that real-time monitoring of landfill gas generated by fermentation of wastes can be performed by in-situ measurements of a portable sensor-system, including wireless functionalities of remote control.

Additionally, odour quantification in a municipal waste site has been validated by off-line conventional olfactometric measurements (regulations EN 13725/2003). A comparison of the sampled air at the waste site between sensor-system and olfactometric panel test of expert assessors has been performed as well. The results demonstrate that the sensor-system is a complementary valid tool to realize a low-cost sensor node in a distributed network for real-time and in-situ odour air-monitoring applications to assess potential annoyance zones.

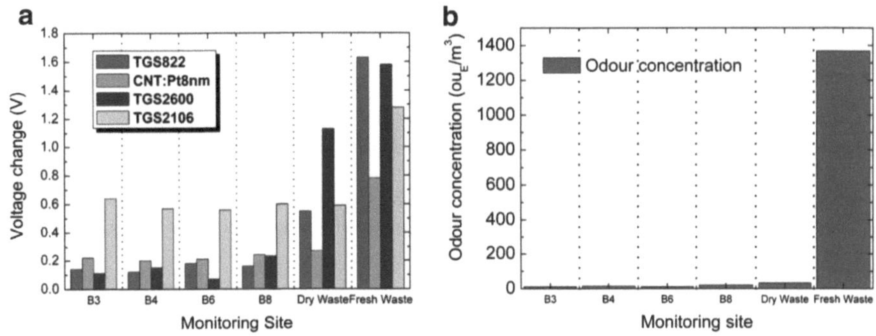

Fig. 15.5 (**a**) Chemical pattern measured by a sensor array based on four elements used to assess malodours sampled in six different bags at a municipal waste site. (**b**) Results of the olfactometric panel test for related corresponding six bags sampled at a waste site

Figure 15.5 shows the chemical pattern of a four-sensor array for six air samples collected in different positions of a municipal solid waste landfill. The *Fresh waste* exhibits the larger response for all sensors. In addition, a comparison of the olfactometric panel test, based on four expert assessors, has been realized for six bags containing malodours sampled at the waste site. The results demonstrate a good overlapping of the chemical patterns between sensors response and olfactometry method. This enables real-time odour monitoring.

4 Conclusions

A portable gas sensor-system based on low-cost n-type and p-type sensing elements has been developed for air-quality control, landfill gas monitoring, and malodours control emitted from a municipal waste site. This portable gas sensor-system has been validated in a real-scenario by chemical analyzers at an air-quality control site (ARPA-PUGLIA), a specific chemical analyzer used in a municipal waste site, and olfactometry off-line method by a test panel, respectively. The results achieved demonstrate that the developed portable gas sensor-system can be used for a new gas sensing paradigm of environmental measurements with a good accuracy, including advanced functionalities of remote control in a distributed sensor network.

Acknowledgements The authors are indebted to Dr. A. Nocioni from ARPA-PUGLIA for science support in the air quality measurements, and Mr. M. Carrozzo from Project Automation for technical assistance during experimental campaign. ENEA team would like to thank ARPA-PUGLIA for joint-research programme to validate developed portable sensor-systems for air quality control.

References

1. Kularatna N, Sudantha BH (2008) An environmental air pollution monitoring system based on the IEEE 1451 standard for low cost requirements. IEEE Sensor J 8:415–422
2. Al-Ali AR, Zualkernan I, Aloul F (2010) A mobile GPRS-sensors array for air pollution monitoring. IEEE Sensor J 10:1666–1671
3. Mead MI, Popoola OAM, Stewart G, Hodgson T, McLeaod M, Landshoff P, Hayes M, Calleja M, Jones RL (2010) Proceedings of sensors systems for environmental monitoring conference, Royal Society of Chemistry, London, 14 Oct 2010
4. Persaud KC, Woodyatt NCP, Sneath RW (2008) Proceedings of the 7th IEEE sensors 2008 conference, Lecce, 26–29 Oct 2008
5. Rock F, Barsan N, Weimar U (2008) Electronic nose: current status and future trends. Chem Rev 108:705–725
6. Di Francia G, Burrasca G, De Vito S, Massera E (2009) Energia, Ambiente ed Innovazione 1:76–89
7. Persaud KC, Wareham P, Pisanelli AM, Scorsone E (2005) 'Electronic nose'—new condition monitoring devices for environmental applications. Chem Senses 30:i252–i253
8. Guo D, Zhang D, Li N, Zhang L, Yang J (2010) A novel breath analysis system based on electronic olfaction. IEEE Trans Biomed Eng 57:2753–2763
9. Zampelli S, Elmi I, Ahmed F, Passini M, Cardinali GC, Nicoletti S, Dori L (2004) An electronic nose based on solid state sensor arrays for low-cost indoor air quality monitoring applications. Sensor Actuator B 101:39
10. Di Natale C, Paolesse R, D'Amico A (2007) Metalloporphyrins based artificial olfactory receptors. Sensor Actuator B 121:238–246
11. Dutta R, Morgan D, Baker N, Gardner JW, Hines EL (2005) Identification of *Staphylococcus aureus* infections in hospital environment: electronic nose based approach. Sensor Actuator B 109:355
12. Ryan MA, Zhou H, Buehler MG, Manatt KS, Mowrey VS, Jackson SP, Kisor AK, Shevade AV, Homer ML (2004) Monitoring Space Shuttle air quality using the Jet Propulsion Laboratory electronic nose. IEEE Sensor J 4:337–347

Chapter 16
A Resistive Sensor for Carbon Monoxide Detection

Alexandro Catini, Francesca Dini, Marco Santonico, Eugenio Martinelli, Andrea Gianni, Corrado Di Natale, Arnaldo D'Amico, Roberto Paolesse, and Alberto Secchi

Carbon monoxide is a odorless, colorless and toxic gas. The effects of carbon monoxide exposure may vary greatly from person to person depending on age, overall health and both concentration and duration of exposure. In this paper the detection of carbon monoxide is accomplished through the conductivity changes of a Cobalt-TriPhenylCorrole (CoTPC) layer deposited onto a micro-fabricated substrate. The influence of light in enhancing sensor sensitivity has also been investigated. Results demonstrate the capability of this sensor to detect carbon monoxide below the limits of toxicity.

1 Introduction

Carbon monoxide (CO) is a colorless and odorless gas produced during combustion processes. For this reason it is one of the most important pollutants in urban areas where most of CO is produced by internal combustion engines.

The importance of accurate carbon monoxide detection resides in its high toxicity even at low concentrations: CO binds hemoglobin causing harmful health effects by reducing oxygen delivery to the body's organs (like heart and brain) and tissues [1]. Being related to the oxygenation of living tissues, the effects of carbon monoxide poisoning result in several symptoms ranging from headaches and dizziness to paresis, convulsions and unconsciousness until myocardial ischemia,

A. Catini • F. Dini • M. Santonico • E. Martinelli • A. Gianni • C. Di Natale (✉) • A. D'Amico
Department of Electronic Engineering, University of Rome Tor Vergata, Rome, Italy
e-mail: dinatale@uniroma2.it

R. Paolesse
Department of Chemical Science and Technology, University of Rome Tor Vergata, Rome, Italy

A. Secchi
SELEX Sistemi Integrati S.p.A, Rome, Italy

A. D'Amico et al. (eds.), *Sensors and Microsystems: AISEM 2011 Proceedings*,
Lecture Notes in Electrical Engineering 109, DOI 10.1007/978-1-4614-0935-9_16,

atrial fibrillation, pneumonia, pulmonary edema, acute renal failure and changes in perception of the visual and auditory systems [2, 3].

Different limits of exposure to carbon monoxide are recommended in various operating situations, the National Institute for Occupational Safety and Health (NIOSH, USA) has established a Recommended Exposure Limit (REL) of 3–5 ppm (corresponding to 40 mg/m^3 in standard atmosphere) as an 8 h time-weighted average.

Due to its ubiquity, CO is an endogenous product of human body where it also play a role in several biochemical processes [3]. For this reason, although its chemical structure makes it compatible with olfactory receptors, it probably saturates the receptors making impossible the detection of airborne molecules.

The great reactivity of CO offers many possibilities for a sensor development. The most diffused sensors are those based on metal-oxide semiconductor materials, such as SnO_2, and amperometric sensors formed by a catalytic metal and a solid electrolyte [4]. In both cases, sensors requires a high temperature. Alternative detection can be based on organic materials that can efficiently bind CO at room temperature. Among these materials of great interest are porphyrins, a class of compounds that can mimic the uptake of CO in humans being a porphyrin the CO binding element of hemoglobin. In particular it is well known that porphyrin change color upon the binding of CO. On these bases, a sensor was developed some years ago where a direct relation between sensor signal and physiological damage was demonstrated [5].

In this paper the sensing properties of a porphyrin analog, called corrole, are investigated. Corroles are contracted aromatic ring with respect to the porphyrin, and they are characterized by peculiar properties, such as unusually high N–H acidity and excellent metal-binding properties with rapid exchange kinetics with respect to porphyrins [6]. Herewith, the CO sensitivity of Cobalt-TriPhenylCorrole (CoTPC) are reported, this molecule when deposited as solid state layer is sufficiently conductive to allow the measure of electric resistance as sensor signal [7].

2 Sensor Development

The sensor developed in this work consists of gold–chromium interdigitated fingers covered with a CoTPC layer and enclosed in a measure chamber. Either glass or alumina are used as substrates for the conductive paths.

Figure 16.1 shows the basic structure of the sensor on the alumina substrate. On one side of the substrate, 100 gold–chromium interdigitated fingers are evaporated. The width of the fingers and the gap between them is 30 μm. The overall sensor dimensions are approximately 1 cm with a height of 100 μm.

The sensing layer was prepared using a mixture of 66% of Dioctyl Sebacate plasticizer (DOS), 33% of PolyVinylChloride (PVC), 1% of CoTPC and THF as solvent. This membrane improves the adhesion of the sensing layer on the substrate and allows also the contemporary optical characterization of the sensor system.

Fig. 16.1 Basic structure of the CO sensor system on a alumina substrate

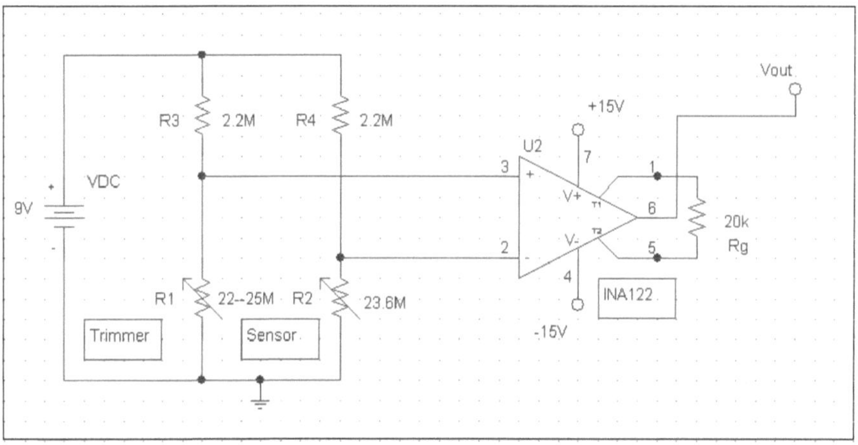

Fig. 16.2 Wheatstone bridge used as conditioning electronics

The sensing layer was deposited onto the interdigitated fingers by drop coating. The conductivity of CoTPC film was measured and it was found to to be sensitive to CO concentration variations. To improve the signal to noise ratio, the sensor was connected in a Wheatstone bridge, as shown in Fig. 16.2.

The sensor in ambient air shows a resistance of 23.6MΩ and a variable resistance of the same order of magnitude was inserted in the opposite bridge branch in order to balance the Wheatstone bridge. A well balanced bridge allows to obtain a higher sensitivity and to significantly reduce the noise. A precision instrumentation amplifier (INA122) was used for accurate signal acquisition.

The system was tested exposing sensor to different concentrations of carbon monoxide (from 5 to 200 ppm) obtained diluting CO in a pure nitrogen flow. The total flux was regulated by a mass-flow controller (MKS Instruments inc.) allowing a known and controlled atmosphere inside the measure chamber.

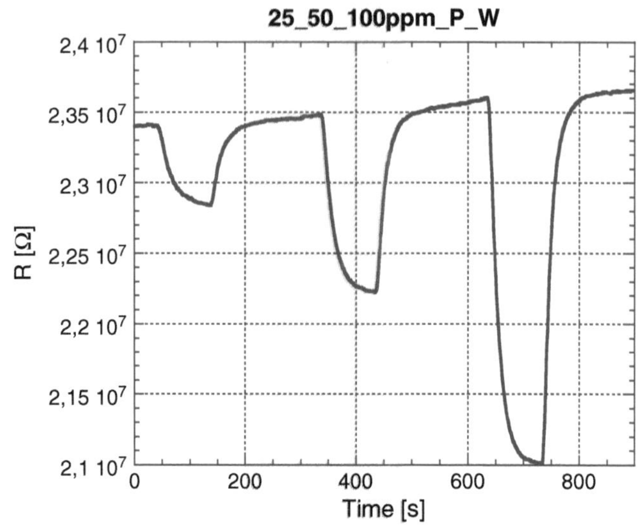

Fig. 16.3 Sensor responses to three different concentrations of carbon monoxide (25–50–100 ppm)

3 Results and Discussion

The sensor response (shown in Fig. 16.3) to three growing CO concentrations is shown as the output resistance value. The sensor shows a good response with an ability to differentiate between small changes in CO concentration and has repeatable response with a fast recovery. Moreover, the detection limit of the sensor allows for the discrimination of the critical threshold of continuous exposure for industrial workers.

Since porphyrinoids are optically active molecule, the sensitivity under illumination was also investigate. For the scope a blue LED was inserted into the measure chamber in order to illuminate the device. The choice of the LED wavelength is due to the CoTPC absorption spectrum peaked at about 443 nm. Preliminary results obtained with this setup show an increase of a factor 2 of the performance with respect to the sensor kept in the dark, reaching a resolution of 5 ppm. Figure 16.4a, b show a comparison between the curves of response of the sensor obtained in dark and under blue light illumination for.

4 Conclusions

In this paper an alternative approach to CO detection has been presented. It is based on the conductivity changes in CoTPC layers. Interestingly, the sensor performance are optically amplified illuminating the sensor with a light corresponding to the main absorption line.

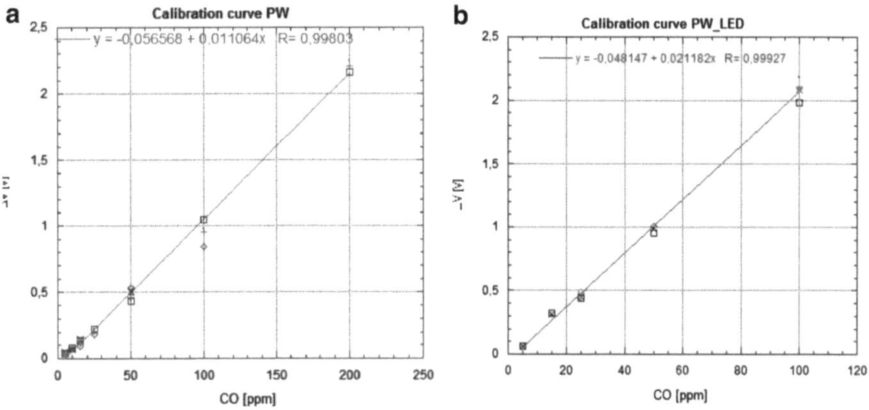

Fig. 16.4 (**a**) Calibration curve of the sensor system without LED illumination. (**b**) Calibration curve of the sensor system with LED illumination

Results demonstrated that this technology is promising for developing low cost and sufficiently sensitive sensors for CO detection in air.

References

1. Henry CR, Satran D, Lindgren B, Adkinson C, Nicholson CI, Henry TD (2006) Myocardial injury and longterm mortality following moderate to severe carbon monoxide poisoning. J Am Med Assoc 295:398–402
2. Ilano AL, Raffin TA (1990) Management of carbon monoxide poisoning. Chest 97:165–169
3. Wu L, Wang R (2005) Carbon monoxide: endogenous production, physiological functions and pharmacological applications. Pharmacol Rev 57:585–630
4. Madou M, Morrison S (1989) Chemical sensing with solid state devices. Academic, Boston
5. Di Natale C, Paolesse R, Alimelli A, Macagnano A, Pennazza G, D'Amico A, Development of porphyrins based sensors to measure the biological damage of carbon monoxide. Proceedings of IEEE Sensors Conference, Toronto, Oct 2003
6. Paolesse R (2008) Corrole: the little big porphyrinoid. Synlett 15:2215–2230
7. Barbe J-M, Canard G, Guilard R (2007) Sensing chemisorption of carbon monoxide by organic–inorganic hybrid materials incorporating cobalt (III) corroles as sensing components. Chem-Eur J 13:2118–2129

Chapter 17
Synthesis, Characterization and Sensing Properties of Nanostructured V_2O_5 Prepared by Electrospinning

V. Modafferi, G. Panzera, A. Donato, P. Antonucci, C. Cannilla, N. Donato, M. Latino, A. Bonavita, and G. Neri

Electrospinning is a simple and inexpensive method for generating nanofibers. In this regard, here we report preliminary data on the morphological and microstructural characterization of V_2O_5/PVAC composite nanofibers prepared by electrospinning from a solution containing poly(vinyl acetate) (PVAC) and vanadium(diisopropr-oxide) in ethanol. The synthesized vanadium pentoxide-based nanofibers have been used as sensing layer for monitoring of low ammonia concentrations in air.

1 Introduction

Metal oxide nanostructures are becoming valuable materials in several applications owing to their surface and size dependant properties. Sensing applications benefit largely of the small particle size and low dimensional structure of these materials.

V. Modafferi • G. Panzera • A. Donato • P. Antonucci
Università "Mediterranea" di Reggio Calabria – Facoltà di Ingegneria – Reggio Calabria, Reggio Calabria, Italy

C. Cannilla
CNR-TAE "Nicola Giordano", Messina, Italy

N. Donato
Dipartimento di Fisica della Materia e Ingegneria Elettronica
Università di Messina, Messina, Italy

M. Latino
Department of Chemical Science and Technologies, University of Rome Tor Vergata, Rome, Italy

A. Bonavita • G. Neri (✉)
Dipartimento di Chimica Industriale e Ingegneria dei Materiali
Università di Messina, Messina, Italy
e-mail: gneri@unime.it

A. D'Amico et al. (eds.), *Sensors and Microsystems*,
Lecture Notes in Electrical Engineering 109, DOI 10.1007/978-1-4614-0935-9_17,

Among the various methods used for the synthesis of nanostructured materials, electrospinning is recognized as a simple and inexpensive technique for generating fibers of various types including polymers, polymer/inorganic composites, and inorganic materials [1].

In this paper we report data on the morphological and microstructural characterization of V_2O_5/PVAC composite fibers prepared by electrospinning. An annealing treatment at high temperatures was performed to obtain the pure V_2O_5 phase. The fibers were deposited on an interdigited ceramic substrate and the sensing properties of devices so fabricated have been evaluated for the monitoring of low ammonia gas concentrations.

2 Experimental

V_2O_5/PVAC composite fibers were prepared by electrospinning, from a solution containing poly(vinyl acetate) (PVAC) and vanadium(diisoproproxide) in ethanol. First, the electrospinning solution was prepared by adding the vanadium oxide isopropoxide to dry ethanol and subsequently to distilled water under agitation, and by stirring continuously for 20 h. Then, the vanadium sol solution was mixed with PVAC under stirring for 5 h to give the final V_2O_5/PVAC solution. The V_2O_5/PVAC composite sol were placed in a syringe and delivered at a constant flow rate using a metallic capillary connected to an high voltage electrical generator. A grounded copper foil, served as counter electrode. When a high voltage was applied a dense mat of V_2O_5/PVAC composite fibers were collected on the copper foil.

The as-spun V_2O_5/PVAC composite was subsequently treated at high temperatures (in the range 300°C–500°C) in air for 5 h, in order to obtain the pure V_2O_5 phase.

Chemoresistive sensors were fabricated depositing by screen-printing thick films of V_2O_5-based materials on alumina substrates provided with interdigited electrodes. The spacing between the Pt electrodes measures 200 μm. Gas sensing tests were carried out inside a stainless-steel chamber under controlled atmosphere. Mass flow controllers were used to adjust desired concentrations of air in nitrogen. The sensors response was measured as change in resistance in four point mode using an Agilent 34970A multimeter. An Agilent E3632A dual-channel power supply was used for the heater of the sensor. The sensors response to ammonia, S, is defined as $S = [(R_{air} - R)/R_{air}] \times 100$ where R is the electrical resistance of the sensor at different ammonia concentrations in air and Rair the baseline resistance in dry air.

3 Results and Discussion

3.1 *Characterization*

SEM image of the as-synthesized sample (Fig. 17.1a) shows that is composed of non-women mats of fibers. The surface of fibers is smooth and uniform. The diameter of each fiber is almost uniform in the whole length. The fibers obtained after calcination exhibit a reduced length and diameter in respect to the as-synthesized V_2O_5/PVAC, due to the decomposition of poly(vinyl acetate). This degradation process is more evident increasing the annealing temperature (Fig. 17.1b). According with the TGA measurements (not shown), the decomposition of PVAC in air initiate around 200°C and proceeds with thermal treatment, leading to the to the breakdown of the polymer backbone and the formation of a very porous and rough structure (see inset in Fig. 17.1b).

In Fig. 17.2 are reported the DRIFT spectra of as prepared and calcined samples. Strong bands observed in the profile of as-synthesized electrospun V_2O_5/PVAC sample between 1,000 and 2,000 cm^{-1} can be assigned to bend and stretching frequencies of PVAC.

After calcination at 500°C, these strong signals disappear, whereas features due to vanadium oxide are evident. In particular, the peak at around 1,020 cm^{-1} is due to the $\nu(V = O)$ mode of V_2O_5, correspondent to the terminal oxygen strongly bonded to only one vanadium atom, while the vibration located at 850 cm^{-1} is due to the bridging oxygen with the stretching modes of the $V = O = V$ bonds. XRD analysis has shown that the as-synthesized electrospun V_2O_5/PVAC sample is amorphous, whereas after the annealing treatment the sample is crystalline.

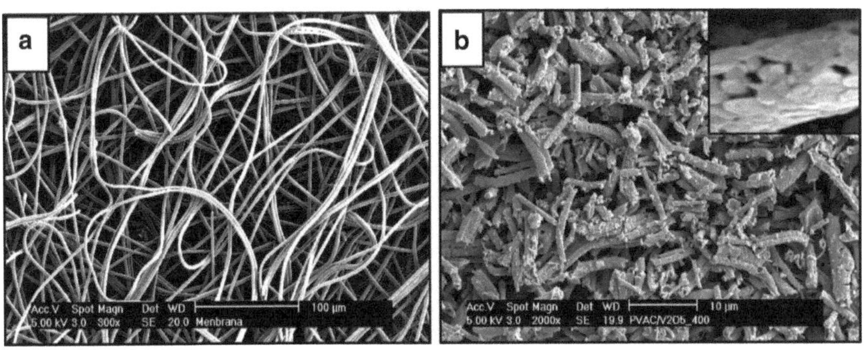

Fig. 17.1 SEM images showing the morphology of V_2O_5 materials prepared by electrospinning. (a) as prepared; (b) after calcination at 400°C. *Inset* shows the rough surface of fibers after calcination at 500°C

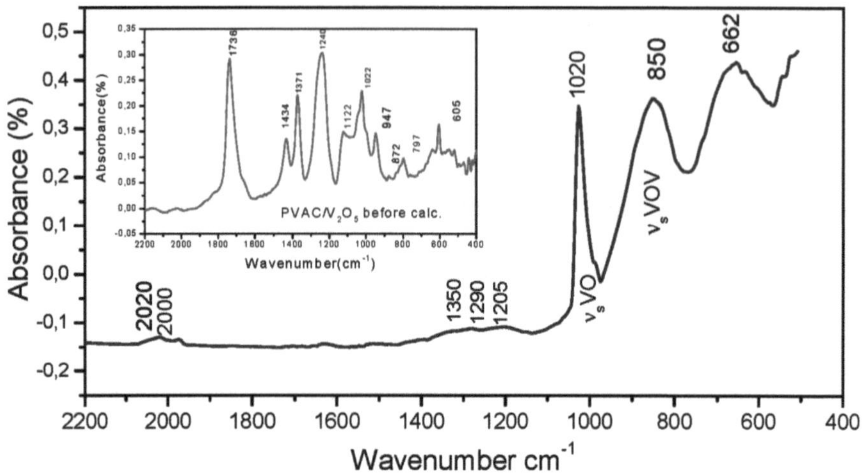

Fig. 17.2 DRIFT spectra of electrospinned samples of V_2O_5/PVAC, as-synthesized (*inset*) and after calcination at 500°C

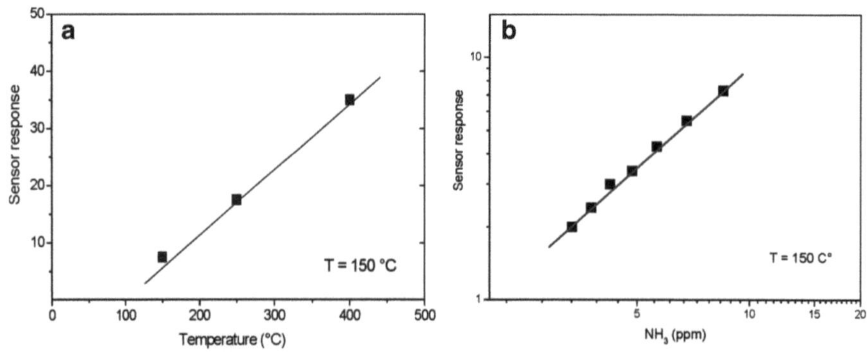

Fig. 17.3 (**a**) Effect of annealing treatment of the as-spun V_2O_5/PVAC composite film on the sensor response. (**b**) Calibration curve of the sensor at low concentrations of ammonia

3.2 Sensing Tests

The sensing behaviour of the samples synthesized has been evaluated in the monitoring of ammonia gas at very low concentrations (from 0.85 to 10 ppm). A preliminary investigation has highlighted that the performance of sensors based on these fibrous nanomaterials depend on the annealing temperature. The effect of annealing treatment of the as-spun composite film on the sensor response to 8.5 ppm of ammonia in air at the operating temperature of 150°C is reported in Fig. 17.3. It can be observed as the thermal treatment improve strongly the sensing properties. This can be correlated, on the basis of characterization studies, with the increase of porosity observed on the annealed samples. A detailed investigation devoted to optimize the performances of sensors based on these fibrous V_2O_5 nanomaterials, is reported elsewhere [2].

4 Conclusion

The synthesis of V_2O_5/PVAC composite nanofibers by electrospinning, and their morphological and microstructural characterization, has been reported. Chemoresistive sensors were fabricated, depositing by screen-printing thick films of the V_2O_5-based materials on alumina substrates provided with interdigitated electrodes. The sensor behavior was investigated in the monitoring of ammonia gas. It has been demonstrated that the developed sensors based on the V_2O_5-based fibrous nanomaterials exhibit high sensitivity to low concentrations of ammonia.

References

1. Huang Z-M, Zhang Y-Z, Kotaki M, Ramakrishna S (2003) A review on polymer nanofibers by electrospinning and their applications in nanocomposites. Composites Sci Technol 63: 2223–2253
2. Modafferi V, Cannilla C, Donato A, Donato N, Spadaro D, Neri G Sol-gel and electrospinning combined synthesis of V_2O_5 fibers for ammonia gas sensors. Sens. Actuators B (submitted)

Chapter 18
Sensing Properties of SnO_2/CNFs Hetero-Junctions

N. Pinna, C. Marichy, M.-G. Willinger, N. Donato, M. Latino, and G. Neri

The sensing properties of SnO_2/CNFs (CNFs = carbon nanofibers) prepared by Atomic Layer Deposition (ALD) have been investigated. By means of a novel ALD approach, which was adapted from the non-hydrolytic sol–gel route, has been possible to achieve the coating of the inner and outer surface of carbon nanofibers with a highly conformal metal oxide film of controllable thickness. The characteristics of oxygen and nitrogen dioxide sensors based on the hybrid nanomaterials have been related to the formation of a p-n heterojunction at the interface between the CNFs and the SnO_2 coating.

1 Introduction

Resistive gas sensors are widely employed in many commercial applications spanning from security and healthcare to environmental and pollution monitoring. The large diffusion in the market is due to their low cost, small dimensions and easy use. In order to develop gas sensors exhibiting improved characteristics, different

N. Pinna • C. Marichy • M.-G. Willinger
Department of Chemistry, CICECO, University of Aveiro, Aveiro, Portugal

N. Donato
Department of Matter Physics and Electronic Engineering
University of Messina, Messina, Italy

M. Latino
Department of Chemical Science and Technologies, University of Rome
Tor Vergata, Rome, Italy

G. Neri (✉)
Department of Industrial Chemistry and Materials Engineering
University of Messina, Messina, Italy
e-mail: gneri@unime.it

A. D'Amico et al. (eds.), *Sensors and Microsystems*,
Lecture Notes in Electrical Engineering 109, DOI 10.1007/978-1-4614-0935-9_18,

routes to the synthesis of new hybrid inorganic/organic nanostructures based on carbon nanotubes have been proposed [1].

In this work, we focused the attention on SnO_2/CNFs (CNFs = carbon nanofibers) nanocomposites prepared by Atomic Layer Deposition (ALD).The ALD approach we used, which was adapted from the non-hydrolytic sol–gel chemistry [2], allowed the coating of the inner and outer surface of CNFs with a highly conformal tin dioxide films of controllable thickness [3].

Tin dioxide deposited by ALD on several supports was already studied as resistive gas sensor [4, 5] but, to the best of our knowledge, there is no works in scientific literature about its use as a coating on CNF for the development of sensing devices.

2 Experimental

The procedure used to prepare SnO_2/CNFs by ALD is reported in detail elsewhere [3]. Tin tert-butoxide and acetic acid were used as metal and oxygen precursors, respectively, of the metal oxide coating. The depositions on the CNFs, previously functionalized by treating them with concentrated HNO_3 at 100°C for 2 h, took place between 100°C and 250°C in an exposure mode reactor. Metal precursor and carboxylic acid were introduced subsequently by pneumatic ALD valves from their respective reservoirs. For the deposition, pure N_2 was used as a carrier gas. The ALD valves opened for 0.03 and 1 s for the oxygen source and tin precursor, respectively. The residence time after each precursor pulse was set to 20 s, followed by a N_2 purge during 15 s.

The sensing device used for testing was consists of an alumina substrate with Pt interdigitated electrodes on one side, and a Pt heater on the other one. The spacing between the Pt electrodes measures 200 microns. The active sensing layer was deposited by screen printing deposition of an aqueous suspension with the coated nanotubes/nanofibers. Gas sensing tests were carried out inside a stainless-steel chamber under controlled atmosphere. Mass flow controllers were used to adjust desired concentrations of air in nitrogen. Electrical measurements were carried out in the temperature range from 50°C to 200°C. The sensors response was measured as change in resistance in four point mode using an Agilent 34970A multimeter.

3 Results and Discussion

The typical morphology of coated SnO_2/CNFs is shown by the TEM micrographs in Fig. 18.1. In these images, the distribution of the metal oxide on the internal and external surface of carbon nanofibers, and the thickness of the coating, can be easily observed.

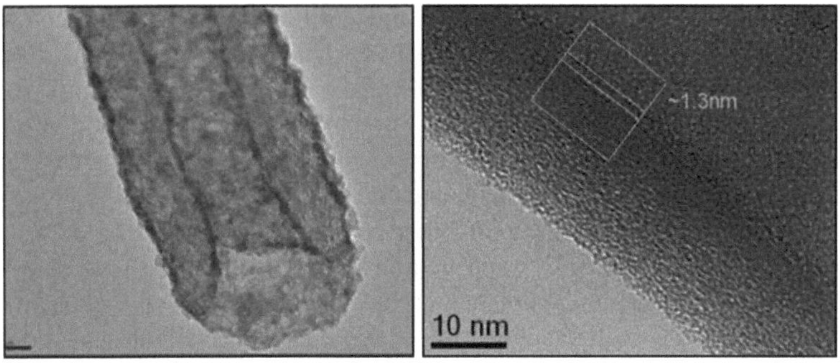

Fig. 18.1 TEM micrographs showing the typical morphology and thickness of coated SnO_2/CNFs

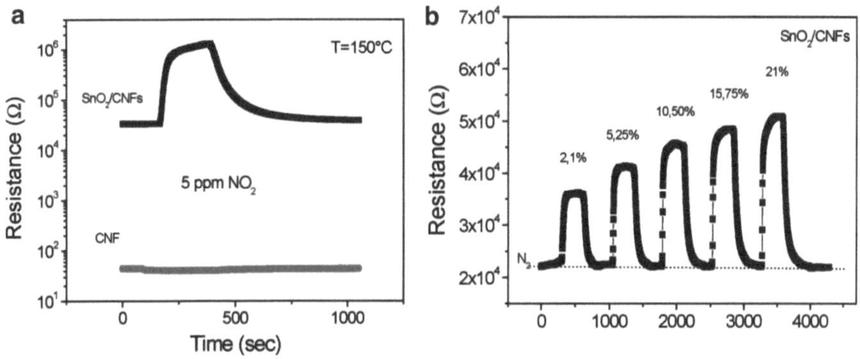

Fig. 18.2 (**a**) Resistance variation of the SnO_2/CNFs sensor to introduction of 5 ppm in air. The response of the bare CNFs sensor is also shown. (**b**) Transient response of the sensor exposed to different concentration of oxygen at 200°C

It appears clear that such a particular distribution of the n-type semiconducting sensing material on the nanotube fibers (having a p-type behavior) can give place to novel electrical properties which could result advantageous for the sensing of gaseous analytes. The electrical and sensing characteristics of chemoresistive devices based on the SnO_2/CNFs heterostructures have been then investigated. Specifically, these sensors were investigated for monitoring of oxygen (O_2) and nitrogen dioxide (NO_2).

Sensing tests revealed that, unlike the bare CNFs, the resistance of a device made of SnO_2-coated CNFs is remarkably and reversibly altered after its exposure to low concentrations of NO_2 and oxygen (see Fig. 18.2a, –b). An increase of the resistance is observed when the sensor is exposed to both oxidizing gases, denoting the n-type behavior of the SnO_2-coated CNFs composite layer.

By changing the ALD process parameters, samples having a different SnO_2 coating thickness have been prepared. An increase of the sensor response with the decreasing of the SnO_2 thickness has been observed. The behavior of the sensors based on these composite materials can be explained by taking into account the electrical modifications induced by the hetero-junction formed at the interface between the n-SnO_2 film and the p-CNFs support. Moreover, the Schottky barrier between the SnO_2 and the CNf is very low since the work function of carbon materials (4.7–4.9 eV) is very close to that of SnO_2 (4.7 eV), making easy for the electrons to travel between the SnO_2 and the CNFs. Then, a modulation of the Schottky barriers (and hence the width of the conduction channel) due to changes in the oxidation state of the SnO_2 (and therefore its work function) accompanying oxygen/nitrogen dioxide adsorption and desorption, can occurs. In such a situation, even a small change in the number of charge carriers available can be easily detected as a resistance change.

4 Conclusion

SnO_2/CNFs nanocomposite materials have been prepared depositing a very thin tin oxide coating layer by means of ALD on the internal and external surface of carbon nanofibers. Since ALD permits a unique control of film deposition, this characteristics was used to modulate the performances of gas sensors as they are closely related to the thickness and the nanostructure of the active material. The electrical and sensing characteristics of the deposited materials were reported. Unlikely the bare CNT/CNF, the resistance of a device made of SnO_2-coated CNT/CNF is remarkably altered after exposure to O_2 and NO_2. The electrical characteristics of the samples can be understood in terms of electrical modifications induced by the hetero-junction formed at the interface between the n-SnO_2 film and the p-CNFs support.

References

1. Ueda T, Takahashi K, Mitsugi F, Ikegami T (2009) Preparation of single-walled carbon nanotube/TiO_2 hybrid atmospheric gas sensor operated at ambient temperature. Diam Relat Mater 18:493–496
2. Niederberger M, Garnweitner G, Pinna N, Neri G (2005) Nonaqueous routes to crystalline metal oxide nanoparticles: formation mechanisms and applications. Prog Solid State Chem 33:59–70
3. Marichy C, Donato N, Willinger M-G, Latino M, Karpinsky D, Yu S-H, Neri G, Pinna N (2011) Tin dioxide sensing layer grown on tubular nanostructures by a non-aqueous atomic layer deposition process. Adv Funct Mater 21:658–666
4. Du X, Du Y, George SM (2008) CO gas sensing by ultrathin tin oxide films grown by atomic layer deposition using transmission FTIR spectroscopy. J Phys Chem A 112:9211–9219
5. Du X, George SM (2008) Thickness dependence of sensor response for CO gas sensing by tin oxide films grown using atomic layer deposition. Sensor Actuator, B 135:152–160

Chapter 19
Response Towards Humidity of Air Stable FETS Based on Polyhexylthiophene Dispersed in Porous Titania

G. Scandurra, A. Arena, C. Ciofi, G. Saitta, S. Spadaro, F. Barreca, G. Currò, and G. Neri

Titanium dioxide films deposited by spraying colloidal dispersions of TiO_2 nanocrystals prepared by laser ablation are characterized by means of Scanning Electron Microscopy (SEM). The metal oxide films are sprayed on the top of gold drain-source contacts thermally evaporated onto SiO_2 coated p-doped silicon wafers. Organic Field Effect Transistors (OFET) are developed by infiltrating polyhexylthiophene solutions into the TiO_2 film. It is found that the presence of the TiO_2 layer improves the air stability of polyheylthiophene OFETs, even in the presence of moisture. In addition, the porous titania layer has remarkable effects on the way the polyhexylthiophene OFETs respond to humidity.

1 Introduction

In the last two decades there has been a growing research interest towards organic thin film devices having Field Effect Transistor (FET) architecture. Organic FETs exploiting as active materials either small organic molecules (phthalocyanines, aromatic compounds, and thiophene oligomers) [1–3], and polymers including functionalised polythiophenes [4], have been developed and used in low temperature gas sensing applications. However, most of the organic semiconductors used as active materials in O-FET have as a common limit a lack of stability that prevents

G. Scandurra • A. Arena • C. Ciofi • G. Saitta
Dipartimento di Fisica della Materia e Ingegneria Elettronica
Università di Messina, Messina, Italy

S. Spadaro • F. Barreca • G. Currò
Advanced and Nano Materials Research s.r.l., Messina, Italy

G. Neri (✉)
Dipartimento di Chimica Industriale e Ingegneria dei Materiali
Università di Messina, Messina, Italy
e-mail: gneri@unime.it

A. D'Amico et al. (eds.), *Sensors and Microsystems*,
Lecture Notes in Electrical Engineering 109, DOI 10.1007/978-1-4614-0935-9_19,

their use in air. The O-FET devices, in fact, are usually stable when operated under nitrogen atmosphere, but do undergo a progressive degradation of their performances when used in air. Degradation, observed in both n-type and p-type semiconducting oligomers and polymers, is often ascribable to undesired doping due to the interaction between the active material and oxygen, resulting in irreversible increase of the material conductivity and in sensitive reduction of the modulation effect of the gate voltage on the drain current. In this paper we show that the performances in air of regioregular polyhexylthiophene (P3HT, purchased from Aldrich) FETs can be improved by infiltrating the active material into nanocrystalline porous titania networks, deposited on the top of the drain-source contacts.

2 Experimental

Regioregular P3HT (electronic grade) was purchased from Aldrich and used as received. Titanium dioxide crystalline nanoparticles (mainly anatase phase) were produced by laser ablation of a metal titanium target in distilled water, irradiating the target surface with the second harmonic (532 nm) output of a 5 ns pulse duration Nd: YAG laser operating at 10 Hz repetition rate. The ablation time was 30 min at a laser fluence of 1 J cm^{-2}. Compact films consisting of nanocrystalline titanium dioxide (mainly anatase phase) were deposited on silicon wafers by the direct nebulization of a TiO_2 colloidal solution through a fine spray nozzle. The films morphology was investigated by means of SEM measurements, performed using a JEOL 5600LV microscope.

Bottom drain-source type OFETs (a typical device and a schematic view are shown in Fig. 19.1) were developed using doped p-Si wafer having 300 nm SiO_2 gate oxide as substrates, and 5 mm long thermally evaporated Au source-drain contacts, spaced by 100 μm. Porous titanium dioxide layers, about 800 nm thick, were deposited on the top of the source-drain electrodes, and infiltrated with chlorobenzene of P3HT. The output characteristics and the trans-characteristics of the developed OFETs were measured at 19°C, under air moisture and in dry conditions using a HP 4155B analyzer.

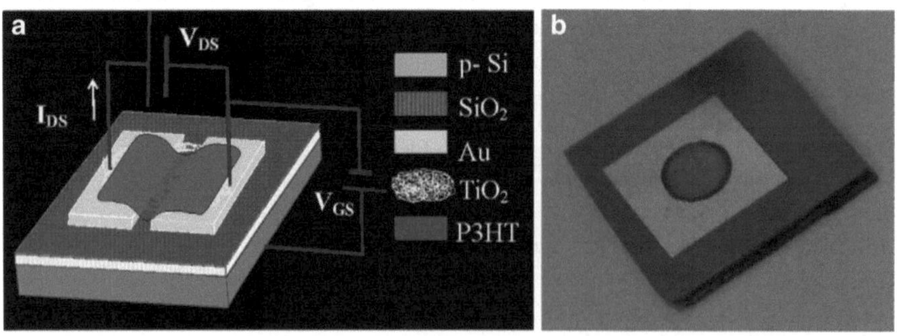

Fig. 19.1 Schematic view (**a**) and magnified image (**b**) of a typical OFET based on P3HT infiltrated into titanium dioxide

3 Results and Discussion

The SEM micrographs of Fig. 19.2a, b show at different magnification the surface of a titanium dioxide film deposited on a silicon substrate. The film is found to be highly porous. Both the "craters" like structure evidenced in Fig. 19.2a and the "fractal" morphology observed in Fig. 19.2b likely arise from the rate at which evaporation of the dispersion medium (water) takes place once the sprayed droplets reach the substrate. Samples morphology, and more generally porosity, could be therefore adjusted by acting on the substrate temperature.

The use of organic semiconductors in OFETs is often limited because of a lack of stability in air. O-FET devices, in fact, are usually stable when operated under nitrogen atmosphere, but do undergo a progressive degradation of their performances when used in air, at high relative humidity (RH). Degradation, observed in both n-type and p-type organic semiconductors, is often ascribable to undesired doping due to the interaction between the active material and oxygen, resulting in irreversible increase of the material conductivity and in sensitive reduction of the modulation effect of the gate voltage V_{GS} on the drain current I_D. Such a poor modulation is clearly evidenced by the characteristics shown in Fig. 19.3a. The source and drain currents I_D and I_S of Fig. 19.3a are measured on a P3HT based FET, having no TiO_2 layer, aged in air for months. Measurements are performed at constant V_{GS}, and plotted against the drain-source voltage V_{DS}. It can be noticed that the maximum relative change of I_D (solid) and I_S (dotted), as V_{GS} changes from −50 V to +50 with 25 V steps, is of the order of 10% only. Figure 19.3b shows the characteristics measured using the same parameters used in Fig. 19.3a, on a OFET aged in air for months, having as active medium P3HT infiltrated into porous titania. Comparison between Fig. 19.3a, b evidences that the presence of titania lowers the current by one order of magnitude, but improves the sensitivity to gate voltage changes. Such a finding could be explained in terms of

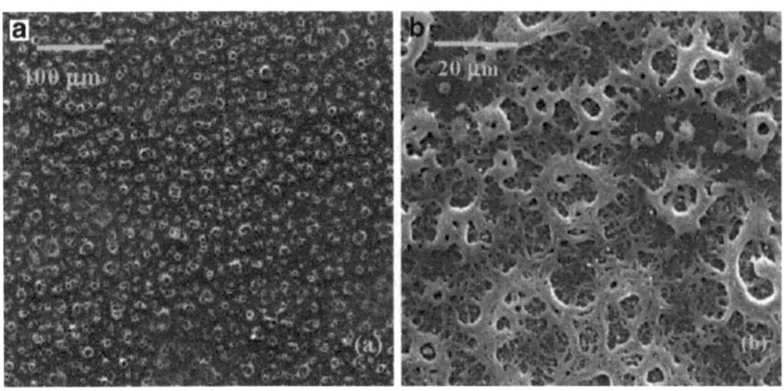

Fig. 19.2 SEM images of a TiO_2 film sprayed on a silicon wafer from a titania colloidal solution

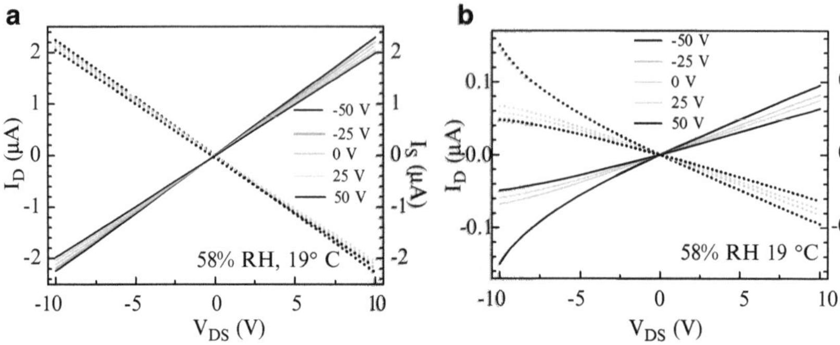

Fig. 19.3 Output characteristics of an O-FET based on P3HT (**a**), and of an OFET having equal design and geometry (**b**), based on P3HT infiltrated into TiO_2. Gate voltages are reported in the *insets*

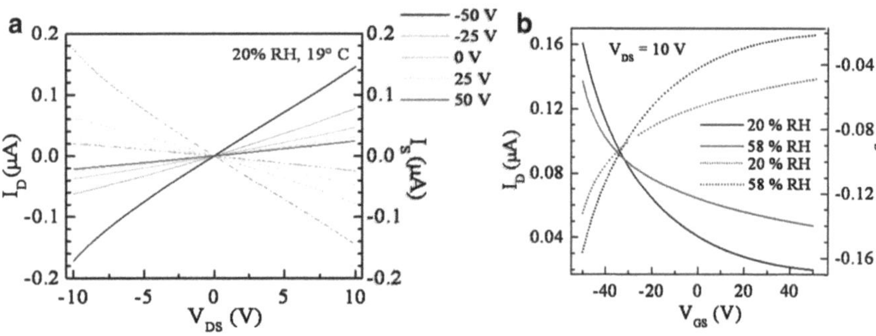

Fig. 19.4 Output characteristics of the same OFET used in Fig. 3b, measured at lower RH (**a**). Transcharacteristics of an OFET based on P3HT/TiO_2, measured at different RH (**b**)

the role played on the local electric field by the interfaces between P3HT and the high dielectric constant TiO_2 layer.

Sensitivity towards V_{GS} of the TiO_2 free FET is found to increase slightly as the RH decreases. Improvements are more sensitive in samples with the titania layer, as it is evidenced by the characteristics measured at 19% RH (Fig. 19.4a) of the same P3HT/TiO_2 OFET used for the measurements shown in Fig. 19.3b.

Interestingly, it is found that the response of the P3HT/TiO_2 based OFETs towards RH, in particular the way the drain and source currents are affected by the RH level, changes with V_{GS}. In fact, it seems that while the absolute value of the current at high negative V_{GS} increases as the RH level decreases, the opposite trend is observed at high positive V_{GS}. Analysis of the transcharacteristics of a P3HT/ TiO_2 FET measured at two different RH levels (Fig. 19.4b) brings to the same conclusion.

4 Conclusions

The presence of a titania layer into the semiconducting material improves the characteristic of OFET based on P3HT operating in air. It seems that exposure to air humidity, known to accelerate OFETs degradation, has reduced impact on OFET based on P3HT/TiO_2 layers, despite the porous TiO_2 films absorb environmental water when exposed to air moisture. Preliminary measurements show that depending on the gate voltage, the drain current of OFET based on P3HT/TiO_2 is found to increase or to decrease with increasing RH.

References

1. Bouvet M (2006) Phthalocyanine-based field-effect transistors as gas sensors. Anal Bioanal Chem 384:366–373
2. Torsi L, Marinelli F, Angione MD, Dell'Aquila A, Cioffi N, De Giglio E, Sabbatini L (2009) Contact effects in organic thin-film transistor sensors. Org Electron 10:233–239
3. Someya T, Dodabalapur A, Huang J, See KC, Katz HE (2010) Chemical and physical sensing by organic field-effect transistors and related devices. Adv Mater 22:3799–3811
4. Fukuda H, Yamagishi Y, Ise M, Takano N (2005) Gas sensing properties of poly-3-hexylthiophene thin film transistors. Sensor Actuator B 108:414–417

Chapter 20
Tuned Sensing Properties of Metal-Modified Carbon-Based Nanostructures Layers for Gas Microsensors

R. Rossi, M. Alvisi, G. Cassano, R. Pentassuglia, D. Dimaio, D. Suriano, E. Serra, E. Piscopiello, V. Pfister, and M. Penza

In this work, carbon nanomaterials have been prepared by CVD technology onto alumina substrates, coated by nanosized Co-catalyst at different thickness (2.5 nm and 7.5 nm) and used for a simple gas sensor device. The surface has been functionalized with sputtered Pt-nanocluster at a tuned loading of 8, 15 and 30 nm. The response of the chemiresistors in terms of p-type electrical conductance has been investigated as a function of the thickness of the Pt-nanoclusters towards different gases (NO_2, NH_3, CO, CH_4, CO_2). Furthermore, the effect of the temperature ranging from 20°C to 250°C on the sensor response has been addressed as well. Additionally, a short-term stability of the carbon nanomaterials based sensor towards NO_2 gas detection has been investigated for a 2-month period. The gas sensors based on Pt-modified carbon nanomaterials exhibit higher sensitivity compared to unmodified material, fast response, reversibility, repeatability, moderate drift of the baseline signal, sub-ppm range detection limit.

1 Introduction

Gas sensors based on carbon-nanostructures (e.g., nanotubes, nanofibers, nanowalls) layers have been largely studied in the form of networked films for highly-sensitive gas detection applications [1–3]. Due to very high surface-to-volume ratio, high electron mobility, great surface reactivities and high capability of gas adsorption, such carbon-based sensing nanomaterials have been investigated as building blocks for fabricating novel devices at nanoscale such as high-performance gas sensors and platforms for biosensing.

R. Rossi • M. Alvisi • G. Cassano • R. Pentassuglia • D. Dimaio
• D. Suriano • E. Serra • E. Piscopiello • V. Pfister • M. Penza (✉)
ENEA, Brindisi Technical Unit for Technologies of Materials, Brindisi, Italy
e-mail: michele.penza@enea.it

A. D'Amico et al. (eds.), *Sensors and Microsystems*,
Lecture Notes in Electrical Engineering 109, DOI 10.1007/978-1-4614-0935-9_20,

However, bare carbon-nanostructures based gas-sensors show low selectivity towards different gases. In order to reduce the power consumption, it is important to increase the sensitivity of the gas sensor at low operating temperatures. Recently, several groups [4–6] have demonstrated that metal nanoparticles deposited on carbon nanotubes networked layers increase sensitivity and selectivity for gas detection, even at low sensor temperatures, due to catalytic effects.

2 Experimental Details

The scheme of the fabricated two-pole chemiresistor is shown in Fig. 20.1. The carbon nanomaterials films were deposited by CVD technology onto cost-effective alumina (5 mm width x 5 mm length x 0.6 mm thickness), coated with growth-catalyst of Cobalt (Co) nanoclusters with a nominal thickness of 2.5 and 7.5 nm and sputtered at 10^{-2} Torr. The Co-catalysed alumina substrates were heated to 550°C by a rate of 10°C/min in H_2 atmosphere upon flow of 100 sccm at a total pressure of 100 Torr. In the gas-plasma, the flow rate ratio between C_2H_2 and H_2 was kept constant at 20/80 sccm, respectively. The film deposition was performed at a constant pressure and temperature of 100 Torr and 550°C, respectively for 30 min by depositing a networked film with a thickness of about 10 μm. In addition, the surface of the carbon-based sensors was functionalized by sputtering of Pt nanoclusters with a loading of 8, 15 and 30 nm for enhanced gas sensitivity and tailored specificity.

All sensors have been located in a test cell (500 mL volume) for gas exposure measurements. The cell is able to host up to four sensors. Dry air was used as reference gas and diluting gas to air-conditioning the sensors. The gas flow rate was controlled by mass flowmeters. The total flow rate per exposure was kept constant at 1,500 mL/min. The sensor temperature was kept constant in the range from 20°C to 250°C. The gas sensing experiments have been performed by measuring the electrical conductance of carbon nanomaterials based films upon controlled

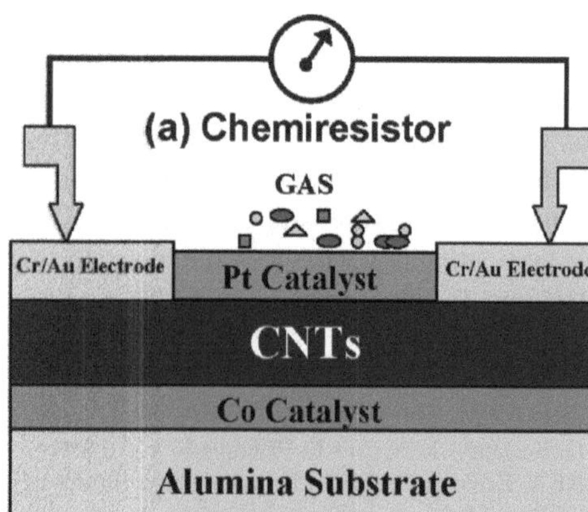

Fig. 20.1 Scheme of the gas sensor based on two-pole chemiresistor using carbon nanomaterials with a laoding of Platinum (Pt) as surface-modification for enhanced gas sensitivity and specificity

concentrations in air of nitrogen dioxide (NO_2), ammonia (NH_3), carbon dioxide (CO_2), methane (CH_4), and carbon monoxide (CO) in the range of 0.01–10 ppm, 10–1,000 ppm, 10–1,200 ppm, 90–5,600 ppm, and 100–1,000 ppm, respectively. The electrical resistance of the sensors was measured by a multimeter (Agilent, 34401A) with a multiplexed read-out by a switch unit (Agilent, 34970A). A J-type thermocouple was used to control the temperature in the test ambient and its output signal was measured by a multimeter (Agilent, 34401A). Data were collected and stored for further analysis in a PC interfaced with a USB/GPIB card in the VEE-software ambient (Agilent).

3 Results and Discussion

The morphology and structure of the carbon based nanostructured films with Pt-functionalizations have been characterized by scanning electron microscopy (SEM), as shown in the Fig. 20.2. A dense network of bundles of carbon nanostructures such as multiple tubes consisting of multi-walled carbon nanotubes, nanostructures like nanofibers, and amorphous carbon appear with a maximum length of 5 μm and single nanostructure diameter varying in the range of 5–35 nm. Nanosized clusters of Platinum (Pt) are clearly visible onto the surface of the carbon-based nanomaterials.

Figure 20.3 shows the repeatability of the response for a 2-month period of two sensors (sensor B and sensor C) in terms of percentage resistance relative change towards three gas concentrations of 10, 2 and 0.5 ppm NO_2, at a sensor temperature of 150°C. These results demonstrate that the repeatability of the gas response for both sensors is acceptable with a variation within a low range.

The measured electrical conductance of the chemiresistor upon exposure of a given oxidizing (NO_2) or reducing (NH_3) gas is modulated by a charge transfer model with *p-type* semiconducting characteristics. Figure 20.4a, c show the typical time response in terms of electrical resistance change for four chemiresistors based

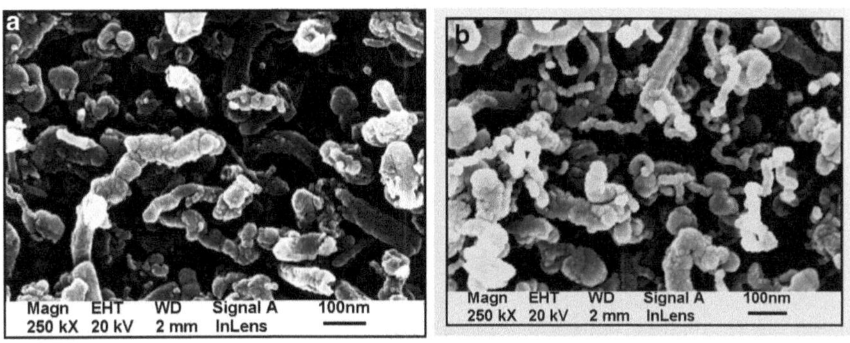

Fig. 20.2 FE-SEM image of carbon nanomaterial film modified with (**a**) Pt 15 nm and (**b**) Pt 30 nm

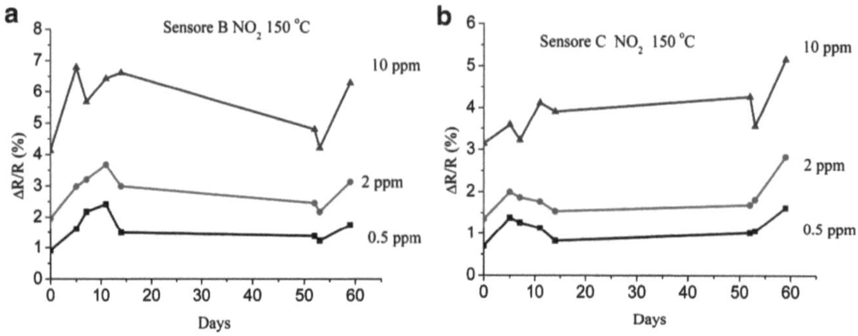

Fig. 20.3 Repeatability of response of carbon nanomaterials sensors to NO_2 for (**a**) Sensor B (Co 2.5 nm, $C_2H_2/H_2 = 20/80$ sccm) and (**b**) Sensor C (Co 7.5 nm, $C_2H_2/H_2 = 20/80$ sccm), at 150°C

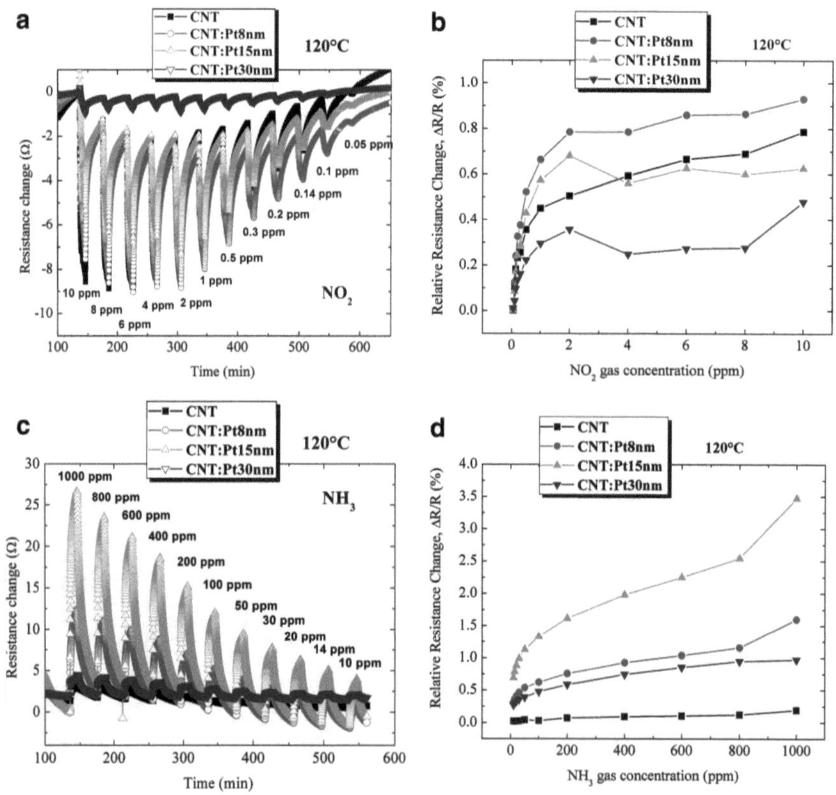

Fig. 20.4 Sensing characteristics of four CNTs-sensors with various Pt loading of 8, 15 and 30 nm towards NO_2 and NH_3 gas, at a sensor temperature of 120°C

on unmodified carbon nanomaterials, and functionalized with a tuned loading of Pt 8 nm, Pt 15 nm and Pt 30 nm, exposed at 120°C to NO_2 and NH_3 gases, respectively. The electrical resistance of all sensors decreases (increases) upon a single gas exposure of the NO_2 oxidizing (NH_3 reducing) gas due to molecules adsorption. In Fig. 20.4b, the chemiresistor functionalized with a Pt-loading of 8 nm shows the highest gas response towards NO_2 at 120°C, while in Fig. 20.4d the chemiresistor with a Pt 15 nm loading shows the highest gas response towards NH_3 at 120°C. As observed in Figs. 20.4b and 20.4d, the thickness of the deposited metal tunes the sensitivity of the gas sensor. Furthermore, for each gas and at a fixed sensor temperature, an optimal thickness of the Pt-metal deposited onto carbon-based nanomaterial is able to maximize the gas response. This can be attributed to a cluster size effect of the catalyst to enhance the gas adsorption by a spillover effect.

4 Conclusions

Carbon nanomaterials based sensors have been prepared by CVD technology with Pt-functionalization at various loading of 8, 15 and 30 nm for enhanced gas sensitivity and tuned specificity to detect NO_2 and NH_3 up to a sub-ppm level of NO_2 gas concentration, and up to a few ppm level of NH_3 gas, at a low temperature of 120°C. Repeatability of the gas response of the unmodified carbon nanomaterials has been successfully perfomed for a 2-month period.

Further measurements are in progress to test the effect of the operating sensor temperature and some process parameters (e.g., precursors gas mixture, Co growth-catalyst size, processing temperature) of the carbon-nanostructures on response of these gas sensors. Finally, these results demonstrate that Pt-modified and carbon-based nanostructures are promising as cost-effective gas sensors for alarm and early detection.

References

1. Kong J, Franklin NR, Zhou C, Chapline MG, Peng S, Cho K, Dai H (2000) Science 287:622–625
2. Someya T, Small J, Kim P, Nuckolls C, Yardley JT (2003) Nano Lett 3(7):877–881
3. Penza M, Cassano G, Rossi R, Alvisi M, Rizzo A, Signore MA, Dikonimos TH, Serra E, Giorgi R (2007) Enhancement of sensitivity in gas chemiresistors based on carbon nanotube surface functionalized with noble metal (Au, Pt) nanoclusters. Appl Phys Lett 90:173123
4. Lu Y, Li J, Han J, Ng H-T, Binder C, Partridge C, Meyyappan M (2004) Room temperature methane detection using palladium loaded single-walled carbon nanotube sensors. Chem Phys Lett. 391:344–348
5. Penza M, Rossi R, Alvisi M, Serra E (2010) Metal-modified and vertically aligned carbon nanotube sensors array for landfill gas monitoring applications. Nanotechnology 21:105501
6. Espinosa EH, Ionescu R, Bittencourt C, Felten A, Erni R, Van Tendeloo G, Pireaux JJ, Llobet EE (2007) Metal-decorated multi-wall carbon nanotubes for low temperature gas sensing. Thin Solid Films 515:8322–8327

Chapter 21
Sub-PPM Nitrogen Dioxide Conductometric Response at Room Temperature by Graphene Flakes Based Layer

Mara Miglietta, Tiziana Polichetti, Ettore Massera, Ivana Nasti, Filiberto Ricciardella, Silvia Romano, and Girolamo Di Francia

The two-dimensional nature of graphene, allowing a total exposure of all its atoms to the adsorbing gas molecules, provides the greatest sensor area per unit volume and outlines the possibility to employ this material as a powerful sensing layer. The synthesis and manipulation of graphene as well as the device fabrication are still challenging due to several technological limits. In the present work we report on a simple approach to fabricate chemiresistive sensors based on chemically exfoliated natural graphite. The devices show the ability to detect a toxic gas, such as NO_2, down to few ppb at room temperature in controlled environments.

1 Introduction

Since the announcement of the isolation of the single layer of graphite, extraordinary efforts have been made to fully explore its manifold and astonishing potential applications. In gas sensing field, the graphene potential has been already investigated finding its ability to detect the presence even of a single interacting molecule [1]. However, until now the fabrication of the single graphene flake based chemical sensor is still challenging due to the complexity of the entire process, starting from the graphene synthesis and/or isolation up to the introduction into the proper device architecture. To date, indeed, several works report on the fabrication of gas sensor devices that employ, as sensing layers, a much more easily manageable material such as the reduced graphene oxide sheets [2–4].

Herein a simple chemiresistor device is described based on chemically exfoliated graphite. Relying on the simple exfoliation methods of the natural graphite reported

M. Miglietta • T. Polichetti • E. Massera (✉) • I. Nasti • F. Ricciardella • S. Romano • G. Di Francia
ENEA Centro Ricerche Portici, Portici (NA), Italy
e-mail: ettore.massera@enea.it

A. D'Amico et al. (eds.), *Sensors and Microsystems*,
Lecture Notes in Electrical Engineering 109, DOI 10.1007/978-1-4614-0935-9_21,

in the literature [5, 6], N,N-dimethylformamide (DMF) and N-methylpyrrolidone (NMP) suspensions of graphene have been prepared and drop-casted onto alumina transducers. So prepared devices have been investigated upon exposure to different gases and the results show a marked and selective response to NO_2 down to few ppb.

2 Experimental

The dispersions have been prepared by bath sonicating 0.5 g of graphite flakes in 50 mL of solvent (DMF, NMP) for 3 h and next separating the larger graphitic particles by centrifugation at 500 rpm for 30 min. The final dispersions are made of flakes which mean size, measured by the Dynamic Light Scattering technique, are 198 nm and 160 nm for the NMP and DMF suspension respectively. Such results are also confirmed by TEM images of the graphene films which show that the suspensions are mainly composed by few layer flakes and sparse thick graphitic fragments (see Fig. 21.1). Few microliters of these dispersions were deposited by drop-casting onto alumina transducers with Au interdigitated electrodes.

The electrical characterization of the sensors was performed using a voltamperometric technique, at constant bias. In such a system (Kenosistec equipment), the device is located in a stainless steel testing chamber placed in a thermostatic box. The testing chamber was provided of an electrical grounded connector for bias and conductance measurements. A constant flow (500 sccm) of carrier gas, i.e., nitrogen or synthetic air was used. The carrier can be properly humidified through a water bubbler placed in a thermostatic bath. At first, the conductance value of the device in its equilibrium state was measured (baseline);

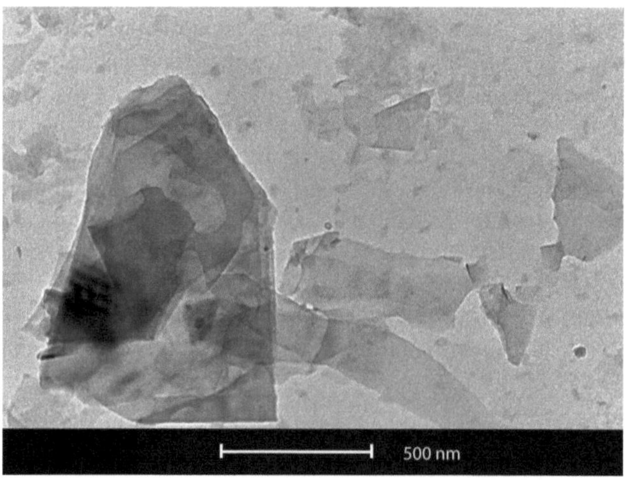

Fig. 21.1 TEM image of graphene flakes obtained by exfoliation with NMP

after that, a given concentration of the analyte can be introduced. Different concentrations of the analyte together with changes in the humidity grade, can be obtained via pneumatic valves and through programmable Mass Flow Controllers. The sensor devices were exposed to a flow of NO_2 and to other analytes: reducing gases such as ammonia, hydrogen, carbon monoxide and oxidant gases such as sulphur dioxide.

3 Results and Discussion

Regardless of the preparation method of the graphene suspension, the sensors show a clear increase of their conductance upon exposure to ppb levels of NO_2 in both dry and wet nitrogen carrier (Fig. 21.2).

As usual for solid state chemical sensors working at room temperature, the conductance exhibits a slow recover to the initial value after the exposure. Actually, as can be seen in Fig. 21.2, wet environment seems to accelerate the desorption kinetics. In fact, the same behaviour can be observed when using a different wet environment such as wet air (Fig. 21.3).

In order to accelerate NO_2 desorption, the device was subjected to a thermal annealing at room pressure and in ambient air [1, 2]. The treatment was effective on the conductance recovery but, on the other hand, the device showed

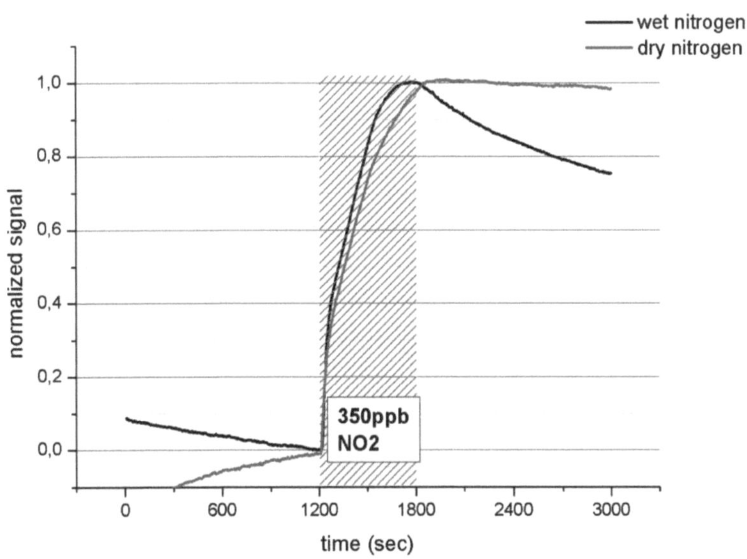

Fig. 21.2 Normalized conductance response kinetics upon exposure to 350 ppb of NO_2 in dry nitrogen and wet nitrogen. Sample is exposed to the analyte in a volume of 0.4 L with a flow of 500 sccm at 22°C. The device is DC biased at 1 V

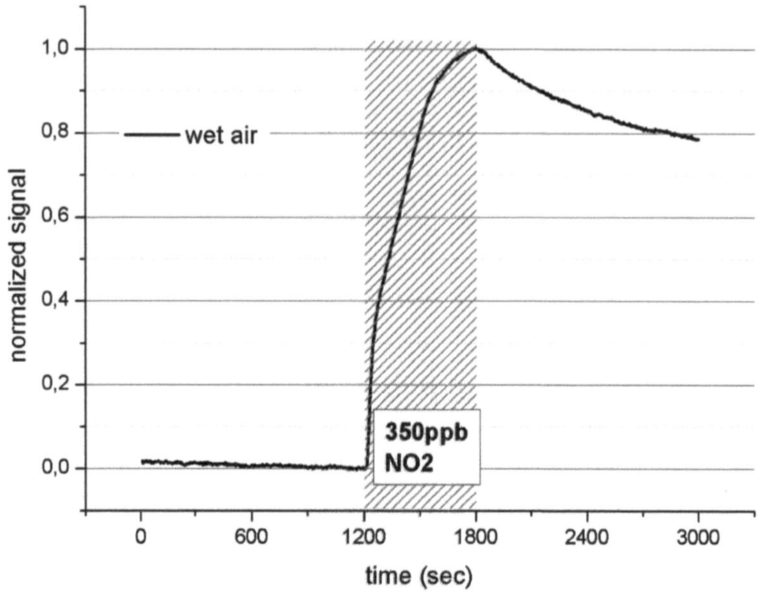

Fig. 21.3 Normalized conductance response kinetics upon exposure to 350 ppb of NO_2 in wet air. Sample is exposed to the analyte in a volume of 0.4 L with a flow of 500 sccm at 22°C. The device is DC biased at 1 V

a marked loss in sensitivity. Actually, conductance recovery within an hour was observed by simply switching-off the applied voltage.

For the other analytes the responses are poorly distinguishable from the electrical noise, hence negligible with respect to the one obtained upon exposure to only 350 ppb of nitrogen dioxide.

4 Conclusions

We have shown that a graphene based chemical sensor can be easily fabricated from chemically exfoliated graphite. The graphene films are sensitive to NO_2 and show, besides, fast response times at room temperature.

The sensor shows a selective response to nitrogen dioxide, with an estimated detection limit as low as 2 ppb, consistent with the best performance observed with few-layers devices. The capacity to achieve such a sensitivity levels can be only ascribed to the nanometric thickness of the sensing layer.

The response and recovery features can help in throwing a light on the interaction mechanism of such a material with the environment, allowing further improvements of the device.

Acknowledgments The authors want to acknowledge Dr. Gennaro Gentile for the TEM measurements. This research was supported by EU within the framework of the project ENCOMB (grant no. 266226).

References

1. Schedin F, Geim AK, Morozov SV, Hill EH, Blake P, Katsnelson MI, Novoselov KS (2007) Detection of individual gas molecules adsorbed on graphene. Nat Mater 6:652–655
2. Dua V, Surwade SP, Ammu S, Agnihotra SR, Jain S, Roberts KE, Park S, Ruoff RS, Manohar SK (2010) All-organic vapor sensor using inkjet-printed reduced graphene oxide. Angew Chem Int 122:2200–2203
3. Robinson JT, Perkins FK, Snow ES, Wei Z, Sheehan PE (2008) Reduced graphene oxide molecular sensors. Nano Lett 8:3137–3140
4. Fowler JD, Allen MJ, Tung VC, Yang Y, Kaner RB, Weiller BH (2009) Practical chemical sensors from chemically derived graphene. ACS Nano 3:301–306
5. Khan U, O'Neill A, Lotya M, De S, Coleman JN (2010) High-concentration solvent exfoliation of graphene. Small 6:864–71
6. Blake P, Brimicombe PD, Nair RR, Booth TJ, Jiang D, Schedin F, Ponomarenko LA, Morozov SV, Gleeson HF, Hill EW, Geim AK, Novoselov KS (2008) Graphene-based liquid crystal device. Nano Lett 8:1704–1708

Chapter 22
Detection of Breath Alcohol Concentration Using a Gas Sensor Array

Gabriele Magna, Marco Santonico, Alexandro Catini, Rosamaria Capuano, Corrado Di Natale, Arnaldo D'Amico, Roberto Paolesse, and Luca Tortora

In this work a system based on an array of five quartz microbalances (QMBs) coated with different metalloporphyrins has been proposed for measuring the Breath Alcohol Concentration (BrAC). Four of these sensors were functionalized with widely selective coatings and one with a metalloporphyrins modified in order to enhanced the sensitivity to alcohols. The results obtained for the BrAC estimation show that the system performances, in terms of accuracy, are absolutely adequate for the legal scopes of breath alcohol measurement.

1 Introduction

The Breath Alcohol Concentration (BrAC) measurement is an in situ and non-invasive procedure to estimate the Blood Alcohol Concentration (BAC). This parameter is mainly used in forensic field by law enforcement agencies as evidential basis for prosecuting drunk drivers. In Italy the legal BAC limit is fixed to 0.5 g/L that corresponds to a BrAC of 125 ppm, using a BAC BrAC Ratio (BBR) of 2,100 [1].

Considering the ethanol sensor development, the measurement of ethylic alcohol exhalation is influenced by several factors with different origins: biological (due to the subject characteristics like: breathing volume, pattern and temperature), procedural (due to the sampling protocol), instrumental (due the type of sensor) and finally due to the process of calibration of the instrument [2]. Moreover in the human breath a wide set of VOCs are present, these compounds could be endogens (acetone, acetaldehyde, isopropanol) or exogenous (toluene, methanol)

G. Magna • M. Santonico • A. Catini • R. Capuano • C. Di Natale (✉) • A. D'Amico
Department of Electronic Engineering, University of Rome "Tor Vergata", Rome, Italy
e-mail: dinatale@uniroma2.it

R. Paolesse • L. Tortora
Department of Chemical Science and Technology, University of Rome
"Tor Vergata", Rome, Italy

A. D'Amico et al. (eds.), *Sensors and Microsystems*,
Lecture Notes in Electrical Engineering 109, DOI 10.1007/978-1-4614-0935-9_22,

and influence the correct BrAC measurement [3–5]. Every commercial device is affected by this kind of problem with the possibility of wrong estimation of BAC values. The current commercial devices are based on three different transduction mechanisms depending on the commercial target: semiconductor, electrochemical and infrared.

In this work an array of QMB-based sensor coated with different metalloporhyrins [6] has been characterized for BrAC measurement.

2 Experimental

An array of five 20 MHz quartz microbalance has been functionalized by spray casting of different metalloporphyrins: one for each QMB. Four of these sensors were coated with Mg-TetraPhenylPorphyrin, Mn- TetraPhenylPorphyrin, Sn-butiloxyl-TetraPhenylPorphyrin, and Zn-butiloxyl-TetraPhenylPorphyrin that are widely selective. In order to enhance the selectivity and sensitivity towards alcohols the array was complemented with Sn-Di-hydro-TetraPhenylPorphyrin. This porphyrin is modified with the addition of OH groups in order to favour hydrogen bonds. For each QMB, the value of each measurement was given by the difference between the resonant frequency measured during the exposure to a reference gas and to the sample (Fig. 22.1)

The system was calibrated exposing sensors to different concentrations of ethanol (from 0 ppm to 180 ppm) obtained mixing the Alcohol vapor at 3000 ppm with a nitrogen flow. The flows of both gases were regulated by a mass-flow controller (MKS Instruments inc.) allowing for a stable and controlled dilution of sample in nitrogen. The humidity influence on the sensibility of the system was evaluated: the

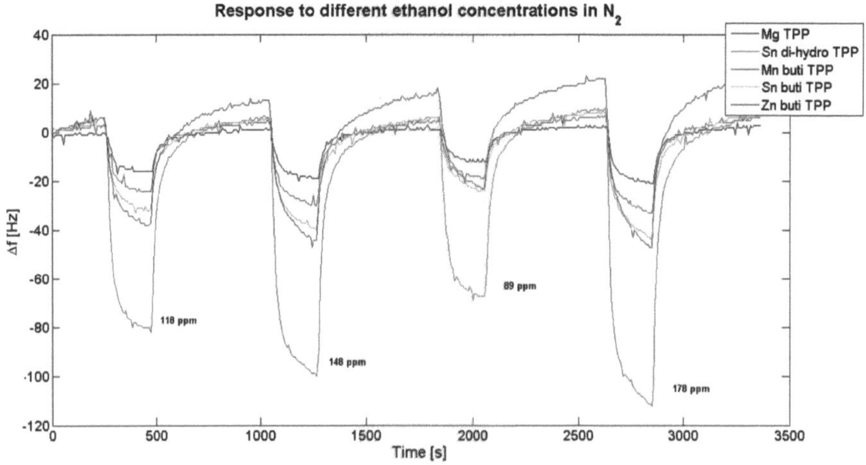

Fig. 22.1 Frequency shifts of each sensor at different ethanol concentration

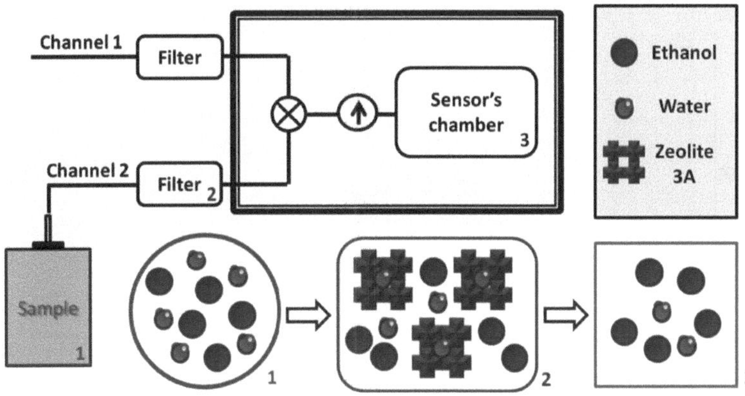

Fig. 22.2 The figure shows the measurement set-up and the mechanisms of water absorption using a zeolite's 3A filter

same ethanol concentrations were considered in a mix with RH fixed at 70%, similar to breath RH.

In order to reproduce the real condition of BrAC measurement the device was tested with breath samples of different person. The sample has been collected in tedlar bags and added with different concentrations of ethanol corresponding to BAC's value from 0.0 to 0.6 g/L. The experimental apparatus was composed of a pneumatic system to transfer the sample from the bag into the sensors cell (Fig. 22.2). Considering that water vapour is one of the principal interference volatile compounds for exhaled breath analysis, a filter of 3 Å zeolites was used to remove it from the sample. The adsorption into zeolites is rule by the pores size, in this case zeolite has a molecular diameter of 3 Å, in order to absorb water (molecular diameter = 2.68 Å) but not ethanol (molecular diameter = 4.46 Å) (Fig. 22.1) [7].

The Fig. 22.3 shows the amount of responses of the different sensors related to different ethanol concentration.

3 Results and Discussion

Sensor array data from breath samples measurement were processed with Partial least square (PLS) in order to retrieve the ethanol concentration. PLS model was properly cross-validated (leave-one-out method) in order to optimize the number of latent variable minimizing the prediction error. Optimized PLS model achieved a Root Mean Square Error of Cross Validation (RMSECV) equal to 0.037 g/L. This value is absolutely adequate for the legal scopes of breath alcohol measurement. Figure 22.4 shows the correlation between predicted and measured BAC.

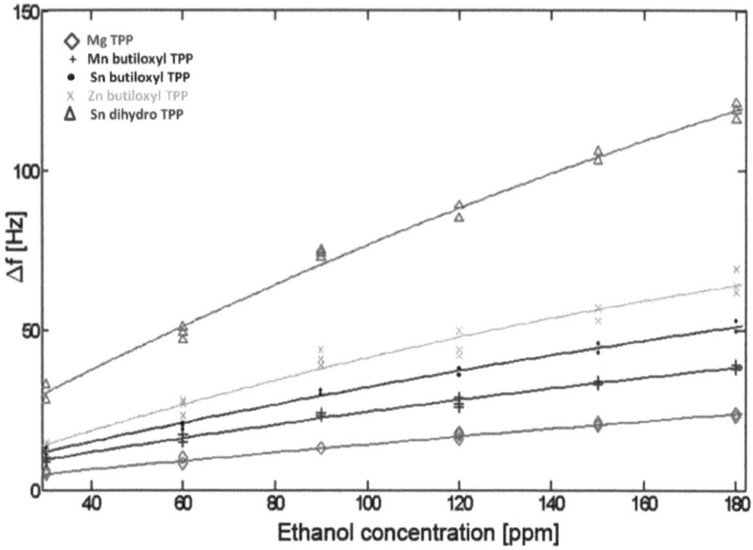

Fig. 22.3 Responses of different sensing materials versus ethanol concentrations

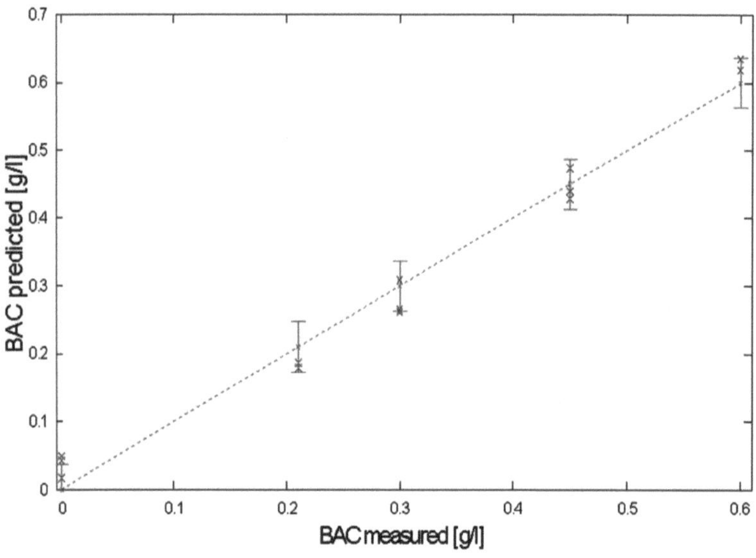

Fig. 22.4 Predicted BAC vs measured BAC

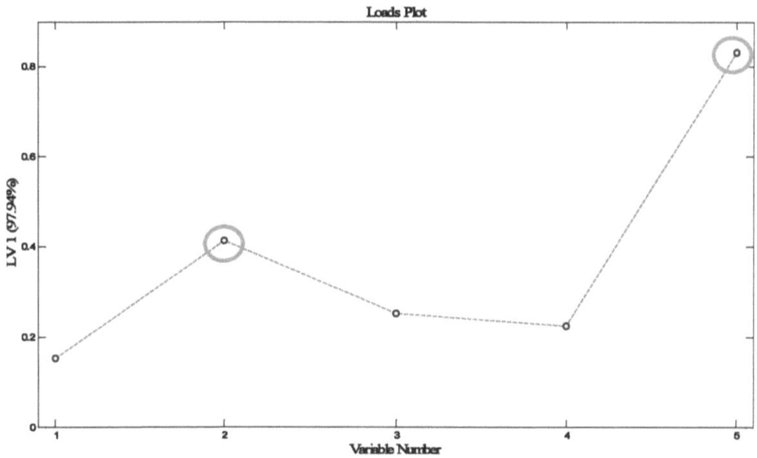

Fig. 22.5 Loadings analysis of PLS model

The analysis of the loadings of the PLS model (Fig. 22.5) gave the evidence that two groups of sensors provide uncorrelated signals, the first is the sensor with enhanced sensitivity to alcohol and the second is formed by the other four sensors. As a confirm of the loading analysis, repeating the PLS analysis with a reduced sensor array formed only by Sn-dihydroTPP and ZnButiTPP provided the same alcohol content estimation accuracy.

4 Conclusions

In this work a device for measuring ethanol exhalation was developed. The complexity of the sample required the study of different aspects. The system calibration was performed in every possible aspect. This allowed us to evaluate the performance of the system in each development phase. The device has shown good performances if compared with commercial breath analyzer.

References

1. Dubowsk KM (1974) Biological aspects of breath-alcohol analysis. Clin Chem 20/2:294–299
2. Gullberg RG (2006) Estimating the measurement uncertainty in forensic breath-alcohol analysis. Accredit Qual Assur 11:562–568
3. Falkenason M, Jones W, SOrbo B (1989) Bedside diagnosis of alcohol intoxication with a pocket-size breath-alcohol device: sampling from unconscious subjects and specificity for ethanol. Clin Chem 35(6):918–921

4. Jones AW, Rossner S (2007) False-positive breath-alcohol test after a ketogenic diet. Int J Obes 31:559–561
5. Shen D, Kang Qi, Wang Yu-E, Hu Q, Du J (2008) New cut angle quartz crystal microbalance with low frequency–temperature coefficients in an aqueous phase. Talanta 76:803–808
6. Di Natale C, Macagnano A, Repole G, Saggio G, D'Amico A, Paolesse R, Boschi T (1998) The explopitation of metalloporphyrins as chemically interactive material in chemical sensors. Mater Sci Eng C Vol 5:209
7. Lalik E, Mirek R, Rakoczy J, Groszek A (2006) Microcalorimetric study of sorption of water and ethanol in zeolites 3A and 5A. Catalysis Today 114:242–247

Chapter 23
Towards a Multiparametric Ammonia Sensor Based on Dirhodium Complexes

S. Lo Schiavo, P. Cardiano, N. Donato, M. Latino, and G. Neri

1 Introduction

Transitions metal complexes (TMCs), exhibiting any physico-chemical change (conductivity, optical, etc.) as a consequence of site specific interactions with gaseous species, can be recognized as a promising source of selective gas sensing materials with enhanced properties compared to those exhibited by conventional metal oxide and polymer gas sensing materials. A major challenge in the development of solid-state chemical sensors is indeed the selective recognition of the target gases. Through weak interaction(s) which selectively binds to a given class of substances, TMCs can give rise to specific, and reversible chemical sensors. Furthermore, such a property can be modulated by an appropriate choice of metal and surrounding ligands.

An additional way to enhance selective recognition is to measure several parameters simultaneously (multiparameter sensor). In this regard, here we investigated the possibility of using several transducer principles in order to improve the gas selectivity of sensors based on dirhodium complexes, by measuring their electrical, optical and acoustic properties.

S. Lo Schiavo • P. Cardiano
Analytical Chemistry and Physical Chemistry, University of Messina, Messina, Italy

N. Donato
Dipartimento di Fisica della Materia e Ingegneria Elettronica, Università di Messina, Messina, Italy

M. Latino
Department of Chemical Science and Technologies, University of Rome Tor Vergata, Rome, Italy

G. Neri (✉)
Dipartimento di Chimica Industriale e Ingegneria dei Materiali
Università di Messina, Messina, Italy
e-mail: gneri@unime.it

A. D'Amico et al. (eds.), *Sensors and Microsystems*,
Lecture Notes in Electrical Engineering 109, DOI 10.1007/978-1-4614-0935-9_23,

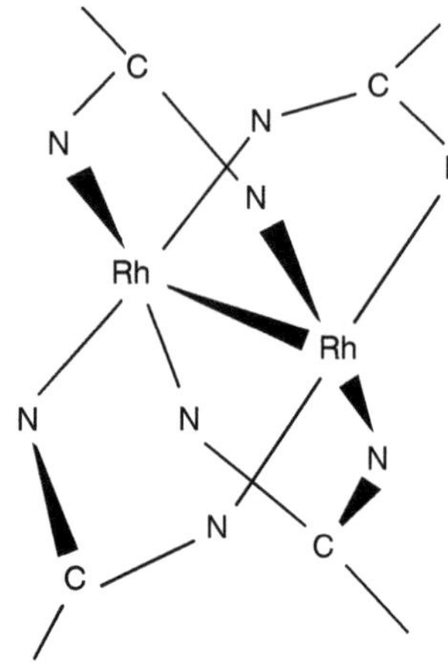

Fig. 23.1 $Rh_2(form)_4$ complex

The dirhodium complexes exhibit a peculiar axial reactivity and are capable to coordinate (reversibly or not) at the axial sites a variety of small molecules (CO, NOx, etc.) as well as O- and N-Lewis bases. Previously, we developed both optical and resistive CO sensors based on these complexes [1–3].

$Rh_2(form)_4$, bearing only formamidinate bridging groups (see Fig. 23.1), has been here investigated as potential sensing material for ammonia detection. In fact, in the presence of formamidinates, better electron-donor ligands than carboxylates previously used, more labile azotate-dirhodium links may be obtained favoring reversible-linking processes during ammonia sensing.

2 Experiments

The synthesis of $Rh_2(form)_4$ has been carried out as previously reported [4]. The complex presents good solubility in many organic solvents and thermal stability in air in the temperature range from RT to 200°C.

Different transducer devices such as optical, resistive, QCM and SAW were fabricated depositing, by simple techniques (drop or spin coating), thin/thick films of $Rh_2(form)_4$ on the sensor substrates. Sensing tests were carried out in the apparatus described in Fig. 23.2.

3 Results

Preliminary results have established that the interaction of thin/thick films of $Rh_2(form)_4$ with ammonia gas resulted in a reversible change of conductance, as well of optical and acoustic properties. In the Figs. 23.3, 23.4 and 23.5 are reported examples of the responses obtained with some of the transducer devices exposed at different ammonia concentrations.

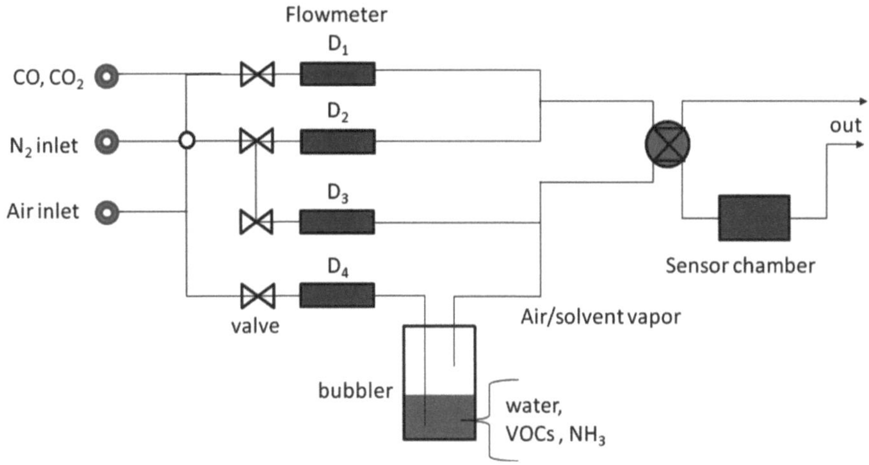

Fig. 23.2 Block diagram of the sensing test apparatus

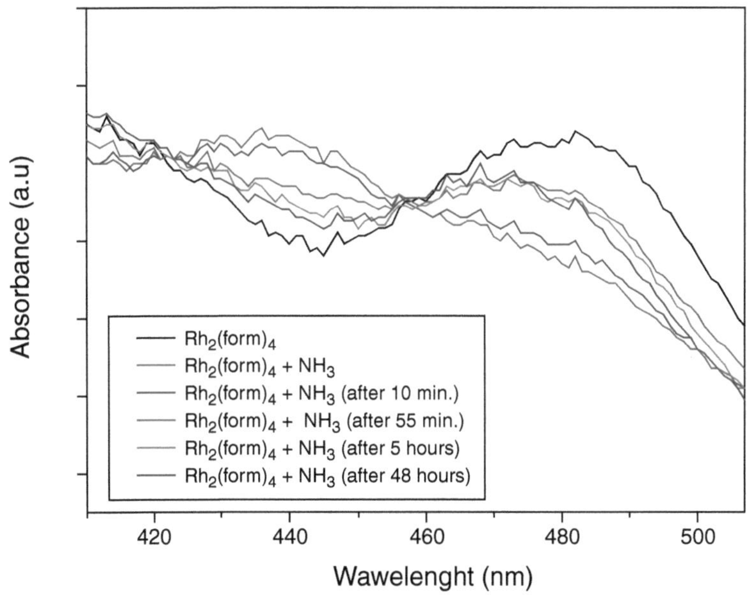

Fig. 23.3 Change of UV-VIS spectrum of $Rh_2(form)_4$ film exposed to NH_3

In Fig. 23.3 is reported the UV-VIS spectrum of $Rh_2(form)_4$ film exposed to ammonia gas. In presence of the gas target, the band at 480 nm associated to $Rh_2(form)_4$ complex disappeared, while a new one was growing at 435 nm, consistent with the formation of a NH_3-dirhodium adduct. The process was reversible and resulted in a well visible change of color, which can be easily monitored.

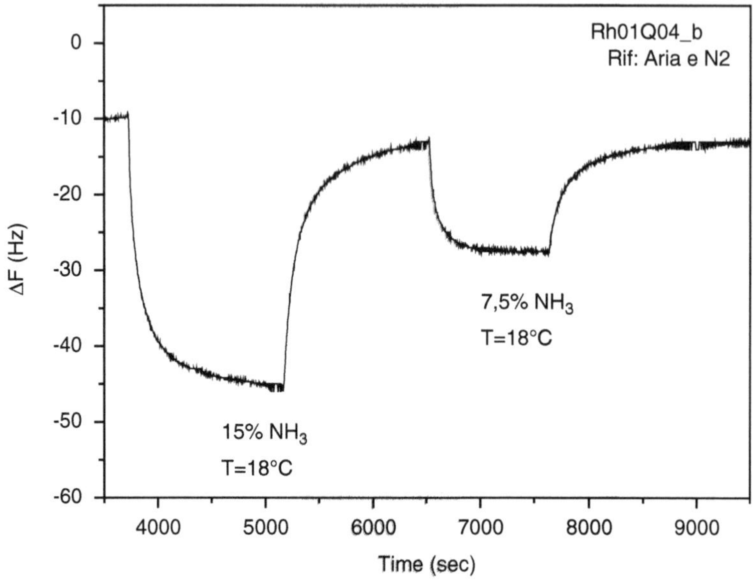

Fig. 23.4 Resonance frequency shift vs. ammonia concentration of QCM device

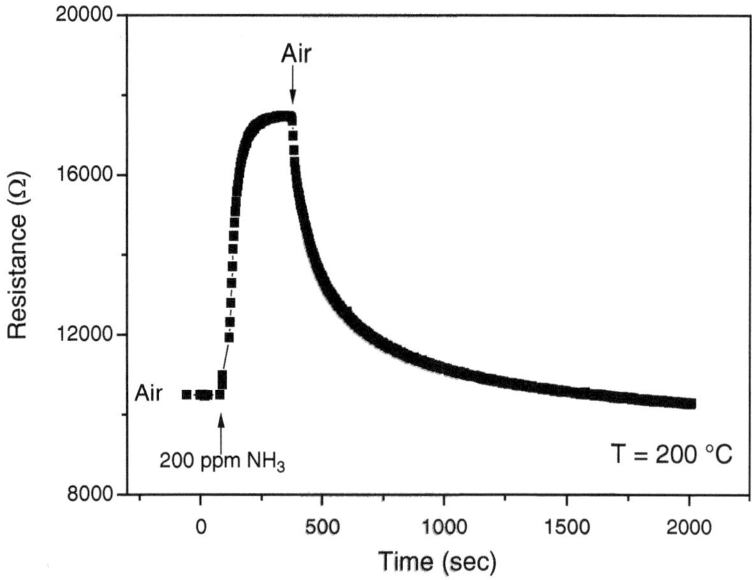

Fig. 23.5 Resistance variation vs. time during an ammonia pulse

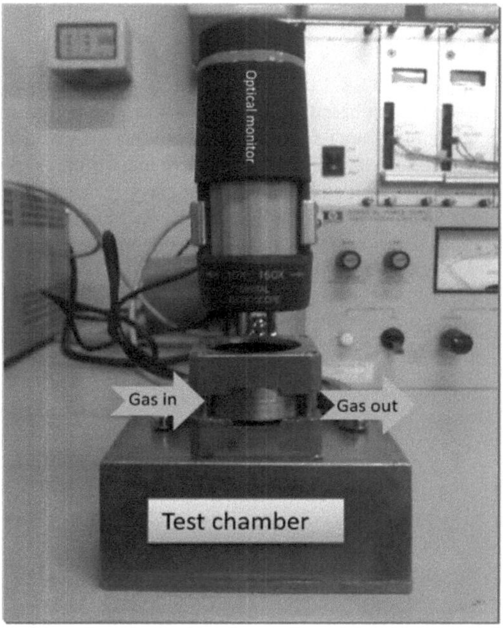

Fig. 23.6 The prototyped solid-state multi-parameter sensor system

In Fig. 23.4 is reported the frequency shift of a QCM microbalance working at 10 MHz coated with the $Rh_2(form)_4$ sensing film, whereas in Fig. 23.5 is described the electrical behavior, at an operating temperature of 200°C, of a resistive sensor fabricated depositing the sensing film on an alumina substrate provided with interdigitated platinum electrodes.

On this basis, we designed and prototyped a solid-state multi-parameter sensor system (Fig. 23.6) based on different sensor platforms, with aim to monitor the target gas by coupling the simultaneous measurement of all parameters.

As a successive development, an appropriate pattern recognition sensing step with aim to further improve results of gas species identification and quantification, will be implemented.

4 Conclusions

In this work are reported the investigation activities about the ammonia sensing properties of $Rh_2(form)_4$ complex. Several transduction mechanisms, by means of optical, resistive, QCM and SAW devices, were investigated. It was well established that the interaction with ammonia gas resulted in a reversible change of conductance, as well of optical and acoustic properties.

Research in progress is aimed to the optimization of the ammonia multiparametric sensor prototype here presented, and its extension to the monitoring of other gaseous species by developing new TMCs sensing materials.

References

1. Hilderbrand SA, Lim MH, Lippard SJ (2004) Dirhodium tetracarboxylate scaffolds as reversible fluorescence-based nitric oxide sensors. J Am Chem Soc 126:4972–4978
2. Gulino A, Gupta T, Altman M, Lo Schiavo S, Mineo PG, Fragalà IL, Evmenenko G, Dutta P, Van der Boom ME (2008) Selective monitoring of parts per million levels of CO by covalently immobilized metal complexes on glass. Chem Comm 2900–2902
3. Lo Schiavo S, Piraino P, Bonavita A, Micali G, Rizzo G, Neri G (2008) A dirhodium(II,II) molecular species as a candidate material for resistive carbon monoxide gas sensors. Sensor Actuator B 129:772–778
4. Rizzi GA, Casarin M, Tondello E, Piraino P, Granozzi G (1987) UV photoelectron spectra and DV-Xα. calculations on diatomic rhodium formamidinate complexes. Inorg Chem 26:3406–3409

Chapter 24
Application of Artificial Neural Networks to a Gas Sensor-Array Database for Environmental Monitoring

L. Trizio, M. Brattoli, G. De Gennaro, D. Suriano, R. Rossi, M. Alvisi, G. Cassano, V. Pfister, and M. Penza

A sensors array based on two different types of chemical sensors such as tin dioxide commercial sensors and carbon nanotubes innovative sensors developed in the ENEA laboratories to monitor gases (e.g., CO, NO_2, SO_2, H_2S and CO_2) of relevance in polluted air has been analyzed. Measurements of chemical sensing of the sensors array have been performed in laboratory to create a database for applying artificial neural networks (ANNs) algorithms to quantify gas concentration of individual air pollutants and binary gas-mixture. A total number of 3,875 data-samples based on 413 distinct gas concentrations measured by 14 gas sensors has been used in the database. The ANN performance has been assessed for each targeted air-pollutant. The lowest normalized mean square error (NMSE) of 6%, 9% and 11% has been achieved for NO_2, SO_2 and CO_2, respectively. In the contrast, NMSE as high as 28% and 39% has been measured for CO and H_2S, respectively. The aim of this study is the selection of an optimal set of gas sensors in the array for enhanced environmental measurements of gas concentration in real-scenario.

1 Introduction

A strong demand of cost-effective and high performance gas sensors involves air quality control [1–4] to preserve public human health and environment. Here, the performance of a gas sensor array is considered to elucidate several aspects of the pattern recognition scheme based on Artificial Neural Networks (ANNs).

L. Trizio • M. Brattoli • G. De Gennaro
Department of Chemistry, University of Bari, Bari, Italy
e-mail: liviatrizio@lenviros.it

D. Suriano • R. Rossi • M. Alvisi • G. Cassano • V. Pfister • M. Penza (✉)
ENEA, Brindisi Technical Unit for Technologies of Materials, Brindisi, Italy
e-mail: michele.penza@enea.it

A. D'Amico et al. (eds.), *Sensors and Microsystems*,
Lecture Notes in Electrical Engineering 109, DOI 10.1007/978-1-4614-0935-9_24,

ANN is an information processing paradigm that is inspired by the way biological nervous systems, such as the brain, process information [5]. The key element of this paradigm is the novel structure of the information processing system. It is composed of a large number of highly interconnected processing elements (neurons) working in unison to solve specific problems. ANNs, like people, learn by example and training. An ANN is configured for a specific application, such as pattern recognition or data classification, through a learning process [6].

In this paper a sensors array based on two different types of chemical sensors has been studied; they are tin dioxide (n-type) commercial sensors (FIGARO, FIS) and innovative carbon nanotubes (p-type) sensors developed in ENEA laboratories to monitor gases of relevance in polluted air. Measurements of chemical sensing of the sensors array have been performed in laboratory to create a database for applying ANNs algorithms to quantify gas concentration of individual air-pollutants and binary gas mixture (NO_2, CO, SO_2, H_2S and CO_2).

The network type used in this paper is the feed forward back propagation. In this case the connections belonging to the first hidden layer are oriented from the input neurons towards the intermediate ones from which the connections towards the output neurons originate. In this kind of network all the connections between neurons of the same level and the signal backward propagation are not allowed; the connections are fundamentally of the forward kind from which the name Feedforward Networks comes.

1.1 Experimental Details

The sensors array consists of 14 gas sensing elements, 1 temperature sensor (National, LM 35), and 1 relative humidity sensor (Honeywell, HIH 3610). The total gas sensors are 14 and listed as 11 commercial sensors (TGS 2602, SP-AQ1, TGS 2100, TGS 2600, SP-AQ2, TGS 2106, TGS 822, SP-31, SB-AQ1, TGS 2600, TGS 4160), and 3 nanotechnology sensors based on carbon nanotubes (CNTs) surface-modified with Au and Pt clusters (CNT, CNT:Pt, CNT:Au). The sensors array studied in the laboratory for gas sensing has been shown in Figs. 24.1 and 24.2. The sensors are configured with output response as voltage change converted from resistance change.

The targeted air pollutants are five gases: CO (10–1,000 ppm), NO_2 (0.05–10 ppm), CO_2 (30–2,000 ppm), SO_2 (0.05–8 ppm), H_2S (0.05–10 ppm) measured at single component and in binary mixture.

For the purpose of the ANNs classifier to determine the gas concentration, five neural networks have been trained with a different number of hidden neurons and individually specialized for each target gas. A total number of 3,875 data samples based on 413 distinct gas concentrations from all 14 gas detectors, one temperature sensor and one relative humidity sensor has been used. Two sets of data are collected for any targeted gas: a training set and a test set. The former set for training of the network, while the test set was used to validate the trained network.

Fig. 24.1 Experimental setup for gas sensing test of sensors array for environmental air monitoring

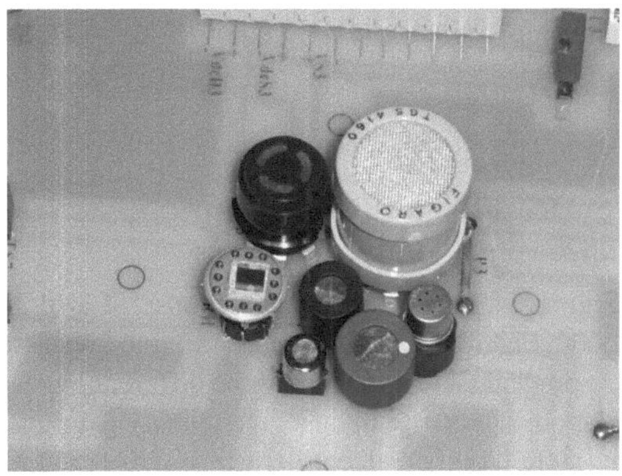

Fig. 24.2 Sensors array based on metal oxides (*MOX*) commercial sensors (Figaro, Fis) and carbon nanotubes (*CNTs*) innovative sensors (ENEA patent)

The data are divided in the following way: CO (training set of 59 data; test set of 14 data), NO_2 (training set of 149 data; test set of 36 data), CO_2 (training set of 30 data; test set of 4 data), SO_2 (training set of 39 data; test set of 8 data), finally H_2S (training set of 67 data; test set of 7 data).

The ANN algorithm used is the common perceptron multi-layer feed-forward network based on error back-propagation [NeuroSolutions 5].

CNTs films for innovative gas sensors were grown by CVD technology. The CNTs films were deposited onto cost-effective alumina (5 mm width × 5 mm length × 0.6 mm thickness), coated with growth-catalyst of Cobalt (Co) nanoclusters with a nominal thickness of 6 nm and sputtered at 10^{-2} Torr. The Co-catalysed alumina substrates were heated to 550°C by a rate of 10°C/min in H_2 atmosphere upon flow of 100 sccm at a total pressure of 100 Torr. In the gas-plasma, the flow rate ratio between C_2H_2 and H_2 was kept constant at 20/80 sccm, respectively. The CNTs deposition was performed at a constant pressure and temperature of 100 Torr and 550°C, respectively for 30 min by depositing a networked CNTs film with a thickness of about 10 μm. In addition, the surface of the CNTs-sensors was functionalized by sputtering of Au and Pt nanoclusters for enhanced gas sensitivity and tailored specificity.

All sensors of the array have been located in five Al-made test cells (100 mL cell volume) for gas exposure measurements. Each PCB case is able to host up to four sensors. Dry air was used as reference gas and diluting gas to air-conditioning the sensors. The gas flow rate was controlled by mass flowmeters. The total flow rate per exposure was kept constant at 1,000 mL/min. The gas sensing experiments have been performed by measuring the electrical resistance of the sensors upon controlled ambient of individual and binary-mixed air-pollutant, at a optimal sensor temperature. The electrical resistance of the sensors was converted into voltage and measured by a multimeter (Agilent, 34401A) with a multiplexed read-out by a switch unit (Agilent, 34970A). A temperature sensor and humidity sensor was used to control the temperature and humidity in the test ambient and its output signal was measured by a multimeter (Agilent, 34401A). Data were collected and stored for further analysis in a PC interfaced with a USB/GPIB card in the VEE-software ambient (Agilent).

2 Results and Discussion

Figure 24.3 shows the time response of a MOX n-type commercial sensor (TGS 2602, Figaro) towards a binary-mixture consisting of NO_2 gas ranging from 0.05 to 9 ppm and CO gas fixed at three different levels of 10, 50 and 100 ppm. The sensor response increases upon a given binary gas pulse. This demonstrates that the binary mixture based on an oxidizing gas (NO_2) and a reducing gas (CO) results globally oxidizing, as proved by increased voltage upon a given binary gas pulse. Additionally, the related calibration curves are shown in Fig. 24.3b. These relationships of

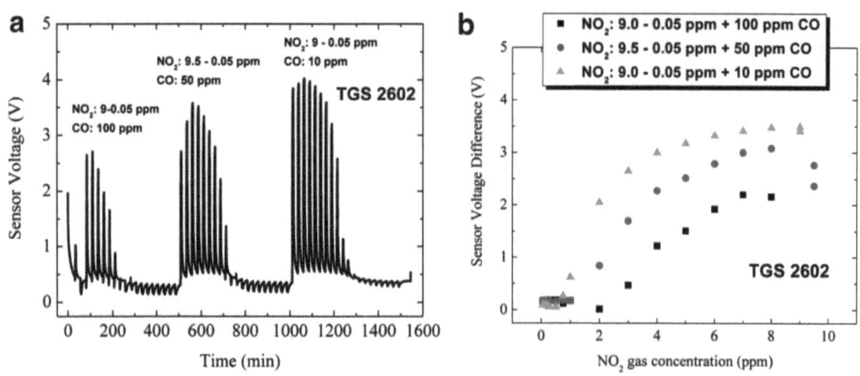

Fig. 24.3 Sensing characteristics of a commercial sensor TGS 2602 (Figaro) towards a binary-mixture of NO_2-CO consisting of NO_2 concentration: 9 (2 expo), 8, 7, 6, 5, 4, 3, 2, 1, 0.75, 0.50, 0.30, 0.25, 0.20, 0.15, 0.10 (2 expo), 0.05 ppm and CO concentration fixed at 100, 50, 10 ppm

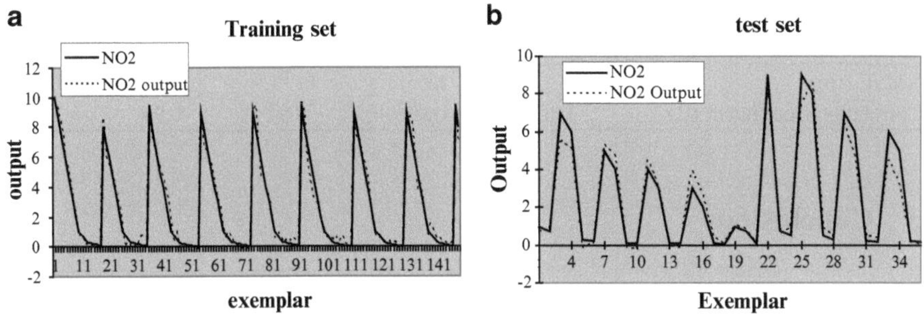

Fig. 24.4 Comparison of real concentration and network output for network (**a**) training and (**b**) test towards NO_2

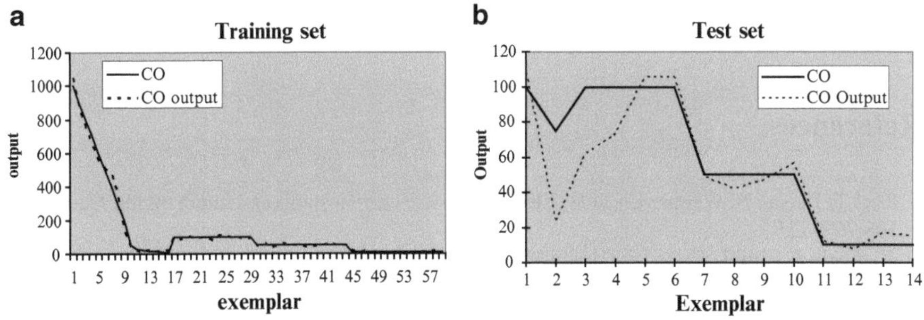

Fig. 24.5 Comparison of real concentration and network output for network (**a**) training and (**b**) test towards CO

voltage change as a function of NO_2 gas concentration for three different CO levels show that the reducing character of the CO gas provides a decrease in the sensor response towards NO_2 with the increasing level of 10, 50 and 100 ppm CO in the binary mixture.

The ANN performance has been assessed. A typical response of the network has been shown for NO_2 and CO gas by using the training set and test set in the Figs. 24.4 and 24.5. The NO_2 and CO characteristics exhibit that real concentrations and predicted concentrations are assessed with a normalized mean square error (NMSE) in the test set as 6% and 28%, respectively. These results demonstrate that the trained ANN is able to determine the NO_2 gas concentration with good accuracy, and lower for CO.

The ANN performance has been also assessed for CO_2, SO_2, H_2S. The lowest NMSE of 9% and 11% has been achieved for SO_2 and CO_2, respectively. In the contrast, NMSE as high as 39% has been measured for H_2S gas. The results achieved are summarised in the Table 24.1. Moreover, the correlation coefficient (R) between real concentration and network output for any gas under test is reported as well.

Table 24.1 Summary of the ANN performance for five gases

Performance test set	CO	NO_2	CO_2	SO_2	H_2S
NMSE (%)	28	6	11	9	39
Correlation Coefficient (R)	0.87	0.97	0.94	0.96	0.86

3 Conclusions

In this paper it has been shown how it is possible to obtain accurate gas detection by a sensors array consisting of commercial n-type MOX sensors and innovative p-type CNT sensors with a low percentage error in concentration estimation by using artificial neural networks. This study addresses to select a set of low-cost gas sensors for environmental air monitoring. Additional measurements in real scenario and dedicated investigations are work in progress.

References

1. Rock F, Barsan N, Weimar U (2008) Electronic nose: current status and future trends. Chem Rev 108:705–725
2. Zampelli S, Elmi I, Ahmed F, Passini M, Cardinali GC, Nicoletti S, Dori L (2004) An electronic nose based on solid state sensor array for low-cost indoor air quality monitoring applications. Sensor Actuator B 101:39
3. Di Natale C, Paolesse R, D'Amico A (2007) Metalloporphyrins based artificial olfactory receptors. Sensor Actuator B 121:238–246
4. Dutta R, Morgan D, Baker N, Gardner JW, Hines EL (2005) Identification of Staphylococcus aureus infections in hospital environment: electronic nose based approach. Sensor Actuator B 109:355
5. Hecht-Nielsen R (1998) Theory of the backpropagation neural network, Proc. IEEE-IJCNN89 at Washinghton DC, vol 1. IEEE Press, New York, pp 543–611
6. Kolehmainen M, Martikainen H, Ruuskanen J (2001) Neural networks and periodic components used in air quality forecasting. Atmos Environ 35:815–825

Chapter 25
Discrimination Between Different Types of Coffee According to Their Country of Origin

Veronica Sberveglieri, Isabella Concina, Matteo Falasconi, Andrea Pulvirenti, and Patrizia Fava

Geographical origin traceability of food is a relevant issue for both producers' business protection and customers' rights safeguard. Between the most widely consumed beverage, coffee is a valuable one, with an aroma constituted by hundreds of volatiles. Differentiation of coffee on the basis of geographical origin still a challenging issue, tough possible by means of chemical techniques. Nonetheless, producer companies need cheaper and simpler tools, able to give a yes/no response in a short time and possibly in a non destructive way. Since the final global volatile composition is also determined by the cultivation climatic conditions, it is in principle possible to distinguish geographical proveniences by exploiting the differences in chemical volatile profile. The present investigation is direct toward the characterization by EN of green and roasted coffees samples according with their geographical origin. The analyzes have been carried out in parallel with chemical classical techniques like GC-MS with SPME. The GC-MS analyses were in good agree with EN results, without sample treatment performed before the analysis.

1 Introduction

The geographical identification of the coffee is of concern felt by both consumers and producers and processors of coffee.

In literature there are several works on the differentiation of coffee on the basis of their botanical species, established through the use of classical chemical techniques [1].

V. Sberveglieri (✉) • A. Pulvirenti • P. Fava
Department Of Agricultural and Food Sciences, Modena and Reggio Emilia University, Reggio Emilia, Italy
e-mail: veronica.sberveglieri@unimore.it

I. Concina • M. Falasconi
CNR-IDASC SENSOR Laboratory and Brescia University, Brescia, Italy

A. D'Amico et al. (eds.), *Sensors and Microsystems*,
Lecture Notes in Electrical Engineering 109, DOI 10.1007/978-1-4614-0935-9_25,

Generally at the time of purchase by the company is performed a visual selection therefore is preliminary concerned with three features:

1 size
2 colors
3 shape

This type of operation is typically accomplished through a visual analysis by expert people, which is in turn the most straightforward, fast and not destructive way to determine coffee origin.

In our work, we aim to differentiate geographical origin of coffee, within the same species of plant (*Coffea Arabica*), both before and after the industrial roasting process.

The roasting process change deeply coffee beans. During this industrial process it happen many chemical reactions, such as the enzymatic browning reaction.

Such as the considerable loss of free water (a_w) and the change in size of coffee beans.

Electronic Nose (EN EOS 835) could be an interesting candidate to assess geographical provenience [2], because it is able to produce rapid response in a not destructive way and usually similar to the human judgment [3].

2 Experimental Section

2.1 Coffee Samples

In this work we have analyzed Arabica coffee samples from South America countries, particularly from Guatemala, Brazil and El Salvador region. Samples from El Salvador came both from the capital and from the countryside.

The coffee according to its area of origin was grow up in area with different altitude.

3 Methods

Electronic noses (ENs) are instruments based on an array of semi selective gas sensors and pattern recognition methods [4]. The EN EOS 835, based on an array of six metal oxide sensors, was equipped with a 40 loading positions auto sampler for static headspace analysis. The analyzes have been carried out in parallel with analytical classical techniques like GC-MS with SPME (Fig. 25.1). Again, have not changed the operation of sample preparation in order to make the results comparable with these two different techniques. In the case of GC-MS was developed a method of tri phases fiber exposure [5].

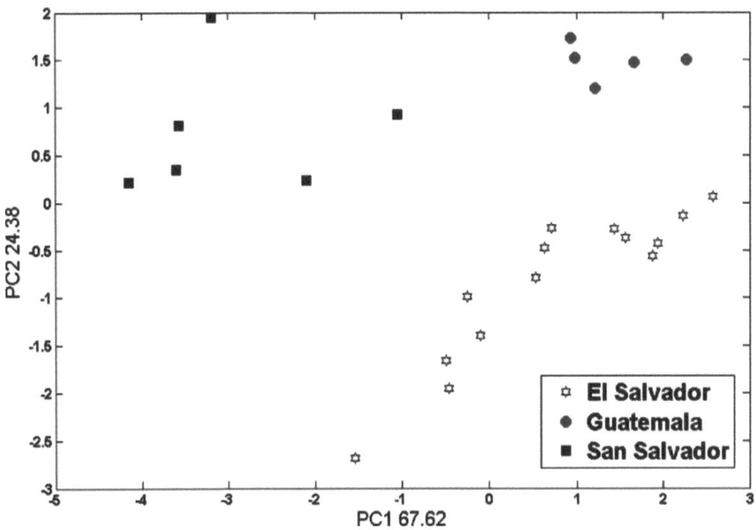

Fig. 25.1 PCA score plot of data related to green coffees: samples grown in different countries and in different region of El Salvador and Guatemala

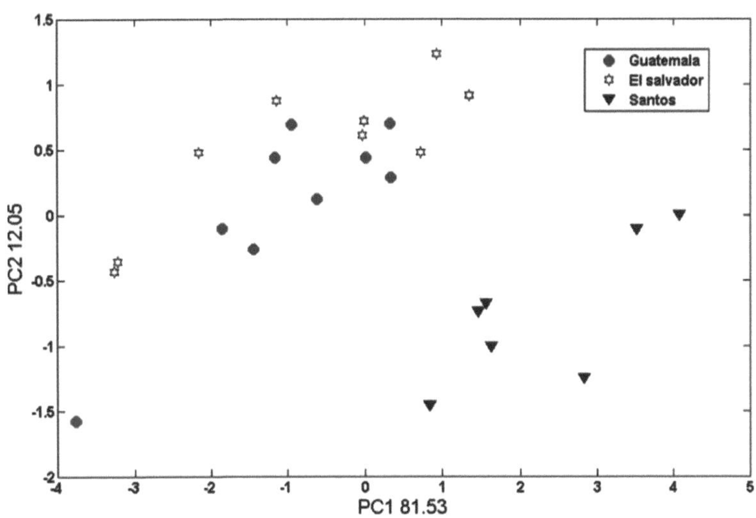

Fig. 25.2 PCA score plot of data related to roasted coffees

4 Results and Discussion

Green coffee samples grown in different locations resulted well separated on principal component (PC) plane.

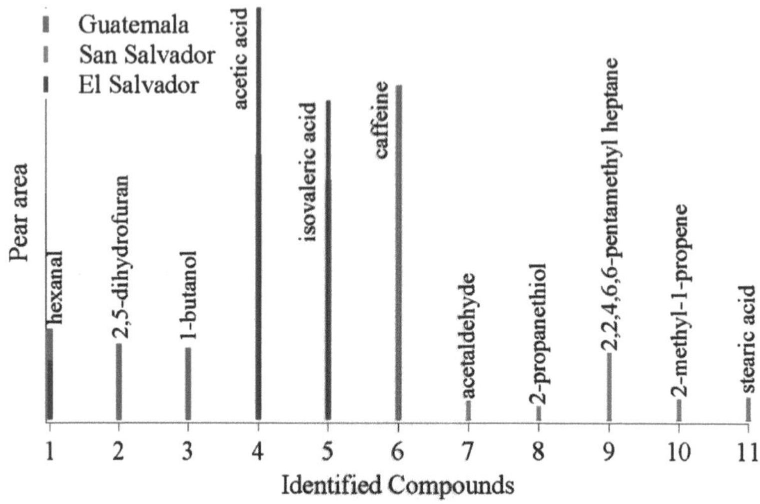

Fig. 25.3 Principal volatile compounds identified by GC-MS with SPME for green coffee sample (Guatemala, San Salvador, El Salvador)

The slight drift along PC1 axis (Fig. 25.2) is attributable to natural rearrangements occurring in the samples over the time, such as oxidative processes. This effect on PC1 has been already observed in literature for food matrices.

Moreover, since all the samples are of *Coffea Arabica* variety, a certain degree of data overlap is predictable.

Nevertheless, EN showed a fair skill in separating and classification test (kNN) indicated a 85% rate of correct classification for roasted coffees. GCMS analyses were in good agree with EN's results: chemical volatile profiles, both for green and roasted samples, evidenced indeed both qualitative (presence of different compounds) and quantitative (relative percentages) differences (Fig. 25.3).

5 Conclusions

Our findings suggest that the EN is both capable to distinguish between coffees grown in different country and also between coffees produced in different region of the same country.

Despite the chemical changes imparted by the roasting process, discrimination was possible both before and after this industrial treatment.

Acknowledgments This work was supported by the FIRB project "Rete Nazionale di Ricerca sulle Nanoscienze ItalNanoNet", Protocollo: RBPR05JH2P, 2009–2013, MIUR

References

1. Alves RC, Casal S, Alves MR, Oliveira MB (2009) Discrimination between Arabica and Robusta coffee species on the basis of their tocopherol profiles. Food Chem 114:295–299
2. Pardo M, Sberveglieri G (2002) Coffee analysis with an electronic nose. IEEE Trans Instrum Measurement 51:1334–1339
3. Pardo M, Don D, Niederjaufner G, Odello L, Faglia G, Sberveglieri G (2000) Discrimination of the certified Italian espresso and prediction of olfactory descriptors with the Pico-1 nose. Proceedings of the 7th international symposium on olfaction & electronic nose, Brighton
4. Cagnasso S, Falasconi M, Previdi MP, Franceschini B, Cavalieri C, Sberveglieri V, Rovere P (2010) Rapid screening of *alyciclobacillus acidoterrestris* spoilage of fruit juices by electronic nose: a confirmation study. J Sensor 2010:9
5. Costa Freitas AM, Parreira C, Vilas-Boas L (2001) Comparison of two SPME fibers for differentiation of coffee by analysis of volatile. Cromatographia 34:647–652

Chapter 26
Evaluation of White Truffle's Aroma with Panelists and a Gas Sensor Array

Giorgio Pennazza, Marco Santonico, Arnaldo D'Amico, Laura Dugo, Chiara Fanal, and Marina Dachà

In this work an array of six quartz microbalances (QMBs) coated with different metalloporphyrins has been used for measuring truffles headspace. The aroma of each truffle has been also tested by a group of expert panelists, scoring it in the range from zero to three. A PLS-DA model built on the QMB array responses have been used to predict panelists evaluations, giving very good results.

1 Introduction

White truffle aroma is strictly dependent on its freshness and extremely important for the determination of the value of the product on the market. A great number of volatile compounds are responsible for truffle aroma, and several chemical reactions can take place during truffle storage, leading to a rapid modification of the aroma itself and of the truffle quality. Aging of truffles following harvesting, in fact, implicates important changes in its flavour, and the indication of a storage technique, allowing to maintain intact the aroma for longer, would be very important.

Two methods can be considered as the gold standard in such a matter: SPME-GC-MS (Solid Phase Micro Extraction – Gas Chromatography – Mass Spectrometry) and expert panelists.

G. Pennazza (✉)
Center for Integrated Research - CIR, Unit of Electronics for sensor systems, "Università Campus Bio-Medico di Roma", via Álvaro del Portillo 21, 00128 Rome, Italy
e-mail: g.pennazza@unicampus.it

M. Santonico • A. D'Amico
Department of Electronic Engineering, University of Rome "Tor Vergata", Rome, Italy

L. Dugo • C. Fanal • M. Dachà
Center for Integrated Research - CIR, Unit of Food and Nutrition, "Università Campus Bio-Medico di Roma", via Álvaro del Portillo 21, 00128 Rome, Italy

A. D'Amico et al. (eds.), *Sensors and Microsystems: AISEM 2011 Proceedings*, Lecture Notes in Electrical Engineering 109, DOI 10.1007/978-1-4614-0935-9_26,

Table 26.1 Table reporting each truffle weight at the beginning of the shelf-life

Truffle	Weight (g)
1	8.4
2	7.02
3	9.68
4	7.71
5	5.44
6	7.32
7	8.29
8	5.92
9	6.52
10	8.44

Recently, arrays of non selective gas sensors, named electronic-noses (e-nose), have been considered as good candidates for food aroma evaluation in many different fields of applications as the monitoring of the shelf life of flavored custard [1] and of fish [2].

Two papers can be found in literature matching electronic nose and the two techniques cited above. One of them proposed a joint-activity of a MOSFET based e-nose and expert panelists in the discrimination of truffles quality [3].

Another work presented a comparison of a MOX (Metal Oxide) e-nose and SPME-GC-MS for the monitoring of truffles aging along a 5 days period [4].

The present work merges these two experiments [3, 4], monitoring truffle aroma modification along a period of 7 days with a QMB (Quartz Micro Balance) based electronic nose and a group of expert panelists.

2 Experimental

An array of six 20 MHz quartz microbalances, covered with six different metallo-porphyrins, has been used to measure the headspace of ten different truffles (see Table 26.1). Samples of white truffle (*Tuber magnatum* Pico) harvested in 2011 during the month of October in Sant'Angelo in Vado (PU, Italy) were subjected to different storage methods and temperatures, to evaluate the variation of the aroma.

The procedure used for the sampling of truffle headspace and its delivery inside the e-nose measure cell is reported in Fig. 26.1.

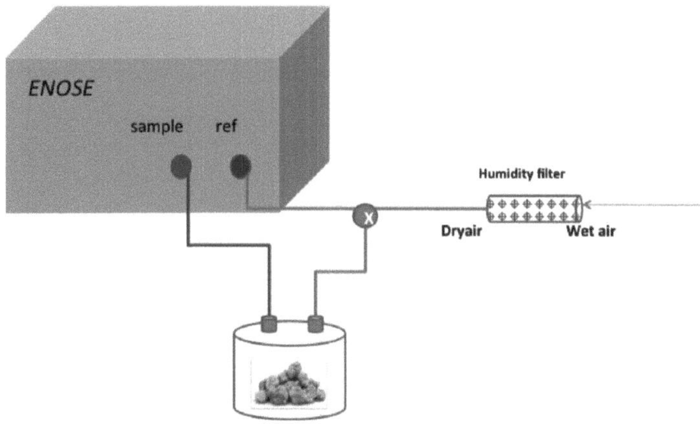

Fig. 26.1 Measurement set-up used to measure truffle headspace with the e-nose

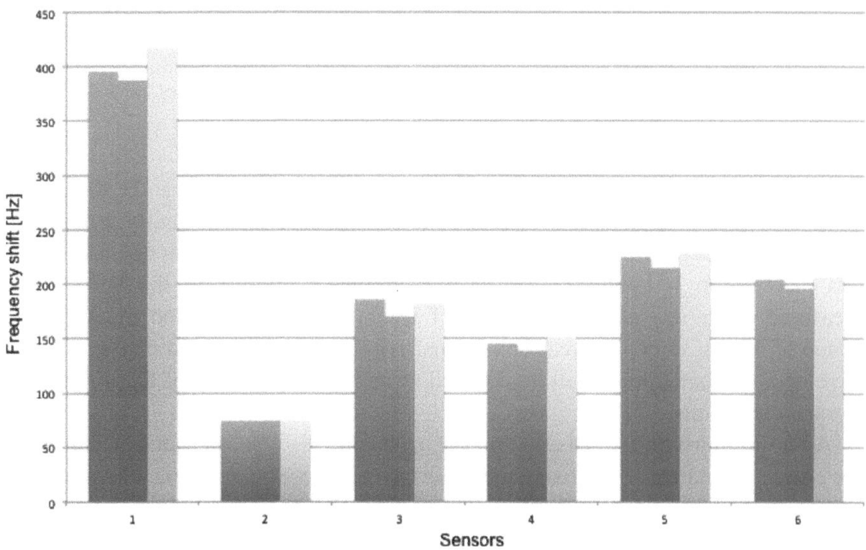

Fig. 26.2 Six-dimensional fingerprint of three of the ten truffles

3 Results and Discussion

Each truffle headspace has given a characteristic fingerprint of six values, calculated as the frequency shifts of each sensor respect to a zero-point referred to the carrier gas. These fingerprints are reported in Fig. 26.2.

From Fig. 26.2 it is possible to observe a common trend of the sensors response for three different truffles of the same species, this suggests that the e-nose is able to identify a volatile signature which is distinctive of the truffle characteristic aroma.

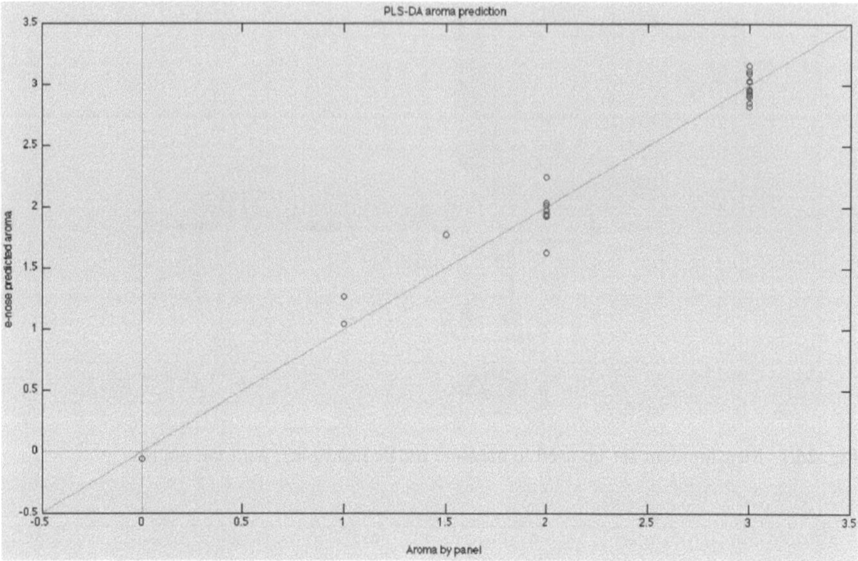

Fig. 26.3 Predicted truffle aroma (e-nose) vs measured truffle aroma (panelist). Scale: 3 (fresh truffle aroma); 2 (less intense truffle aroma); 1 (other aroma); 0 (no aroma)

Besides, the differences in magnitude evidenced for each truffle can be explained by the different weight. Thus, building a model to predict truffle quality, a normalization respect to the truffle weights has been calculated. The dataset considered for this elaboration is composed of 70 measurements relative to the ten truffles measured every day along a period of 1 week. During this period truffles were stored at 4°C in three different conditions: enveloped in rice or in paper; without envelope. Further studies will be devoted to identify the effects of these different storage conditions.

A Partial Least Square Discriminant Analysis (PLS-DA) has been calculated on the e-nose normalized data, using the expert panelist evaluation as the classification scores of the ten measured truffles. This model has been cross-validated via the Leave-One-Out criterion. Figure 26.3 accounts for a high correlation between predicted and measured values, showing a good ability of the e-nose in following the panelists evaluation on the truffles quality. The RMSECV (Root Mean Square Error Cross Validation) of this model is about 0.24, which represents a relative error of 8% respect to the total range of variability.

4 Conclusions

A QMB based gas sensor array with metalloporphyrins as sensing material is able to identify the volatile signature of the white truffle (*Tuber magnatum* Pico) harvested in Sant'Angelo in Vado (PU, Italy), and this fingerprint is dependent on the truffle weight.

The evaluation of truffle quality based on e-nose measurements is similar to the one performed by expert panellists.

A deeper study is necessary to test e-nose capability in monitoring truffle shelf-life and the effects of different storage conditions.

Acknowledgements Thanks to Dr. Luigi Cucchiarini, Professor at the Department of Biochemistry and Molecular Biology of the University of Urbino 'Carlo Bo', for providing truffle samples.

References

1. Santonico M, Pittia P, Pennazza G, Martinelli E, Bernabei M, Paolesse R, D'Amico A, Compagnone D, Di Natale C (2008) Study of the aroma of artificially flavoured custards by chemical sensor array fingerprinting. Sensor Actuator B: Chem 133:345–351
2. Alimelli A, Pennazza G, Santonico M, Paolesse R, Filippini D, D'Amico A, Lundstrom I, Di Natale C (2007) Fish freshness detection by a computer screen photoassisted based gas sensor array. Analytica Chimica Acta 582:320–328
3. Zeppa G, Gerbi V (2001) Using electronic nose" to discriminate white truffle (Tuber magnatum Pico) quality. Sci des Aliments 21(6):683–695
4. Falasconi M, Pardo M, Sberveglieri G, Battistutta F, Piloni M, Zironi R (2005) Study of white truffle aging with SPME-GC-MS and the Pico2-electronic nose. Sensor Actuator B: Chem 106:88–94

Chapter 27
A Semi-Supervised Learning Approach to Artificial Olfaction

Grazia Fattoruso, Saverio De Vito, Matteo Pardo, Francesco Tortorella, and Girolamo Di Francia

In the last decade, semi-supervised learning (SSL) has gained an increasing attention in machine learning. SSL may obtain performance gains by adding to the supervised information, provided by a limited labelled training set, the information content embedded in an unsupervised sample set. This may be very helpful, since obtaining supervised samples can be difficult and costly, as in several artificial olfaction (AO) problems. In this work, co-training style semi-supervised algorithms are applied to air pollution monitoring, an on-field artificial olfaction problem. The primary purpose is to adapt a regressor knowledge to the well known sensors and concept drift issues that characterize the use of solid state chemical sensors in harsh environments. The response of an array of solid state chemical sensors, located in a city street affected by heavy cars traffic, has been monitored for more than 1 year and used to estimate hourly pollutants concentrations. Conventional analyzers provided the needed ground truth. Results obtained by the proposed approach show that it can both reduce the number of labeled samples needed for the multivariate calibration of the device and the performance decay due to drift effects.

G. Fattoruso (✉) • S. De Vito • G. Di Francia
Base Materials and Devices Department, ENEA – National Agency for New Technologies, Energy and Sustainable Development, Portici (NA), Italy
e-mail: grazia.fattoruso@enea.it

M. Pardo
Institute of Applied Mathematics and Information Technology, Genova, Italy

F. Tortorella
Information Engineering Department, Università di Cassino, Cassino (FR), Italy

A. D'Amico et al. (eds.), *Sensors and Microsystems: AISEM 2011 Proceedings*, Lecture Notes in Electrical Engineering 109, DOI 10.1007/978-1-4614-0935-9_27,

1 Introduction

Nowadays, atmospheric pollutants are considered responsible for the increased incidence of respiratory illness in citizens. Some of them (e.g. benzene) are even known to induce cancer in case of prolonged exposure [1]. Actually, pollutants diffusion is heavily affected by atmosphere dynamics and the availability of a limited number of measurement nodes may lead to the misevaluation of the real distribution of gases and particles concentrations in a complex and turbulent environment such as a city [2]. Chemical multi-sensor devices, often referred as electronic noses, are recognized as a possible solution to realize pervasive monitoring of pollutants in city environments. However, their concentration estimation capabilities are seriously limited by the known drift and selectivity issues of solid-state sensors they rely on [3]. Solid state chemical sensors, in facts, are definitely non-selective sensors; their transduction mechanism is not influenced by the primary gas they were designed for but also by the presence of multiple secondary chemicals, known in the chemometrics community as interferents. Sensors drift causes each sensor response to vary during time and it is caused by multiple phenomena like sensor poisoning (e.g. saturation of binding sites, non-reversibility, etc.). In harsh traffic environments, such effects reveal even more evident [4]. On-field applications are furthermore affected by changes in environmental conditions (i.e. T, RH, etc.) and pollutants concentration ratios [5]. It should also be considered that obtaining on-field supervised samples (e.g. a ground truth) may require the simultaneous presence of a mobile station equipped with conventional analyzer for as long as needed to build an adequately representative training set. Hence, this is very costly and definitely unfeasible for a network built up by tenths or even hundreds of electronic noses.

Semi-supervised learning methodologies [6], can represent an interesting solution to reduce the number of labeled samples needed to train the statistical regressor. We also believe that they can be used to reduce sensor drift effects, arising during operative time, that hamper the trained regressor performances.

In this work we present the results obtained by a modified cotraining style semi-supervised regressor used together with a filter style methodology for drift effect and training set dimension reduction in a pollution monitoring on-field application. We show how these methodology can achieve significant improvements over both of the intended goals.

2 On Field Air Pollution Monitoring Dataset

A compact (volume $= 9.7 \times 10^{-3}$ m^3), low cost, solid state multisensor device [5], has been co-located with a conventional air pollution analyzer, operated by Italian Regional Environmental Protection Agency (ARPA). The conventional analyzer response has been used to provide the true concentration values of the target pollutants at the measurement site. These values have been hence used as a reference for the development of a multivariate regression system designed for calibrating the multisensor device response. Conventional fixed station provided concentration estimation for CO (mg/m^3), (g/m^3), C6H6 (g/m^3), NOx (ppb), NO_2 (g/m^3). It was sampled recording hourly averages of the concentration values. The multisensor device was also sampled to provide the hourly average of the resistivity expressed by CO, NO_x, O_3, NMHC and NO_2 targeted metal oxide (MOX) chemiresistors and the commercial temperature and relative humidity sensors. Measurement campaign took place using one of the main street in the centre of an Italian city as testing site; it was characterized by heavy car traffic. Data acquisition campaign lasted from March 2004 until April 2005.

3 Experimental and Methods

In the framework of cooperative semi-supervised learning [7], we tested the performance of two regression-oriented algorithms: the state of the art COREG algorithm [8] and an originally variant developed by us. In previous works, we have shown how a neural network based regressor, adequately designed, trained with a training set including minimum 360 supervised samples was able to achieve optimal sample-by-sample concentration estimation performance. However, the obtained calibration lost precision over time, due to sensors and concept drifts [5, 9]. Actually, the calibration of such a device would be rather a costly process requiring the simultaneous presence of a conventional mobile station providing ground truth measurements for more than 10 days.

In this work, we aimed at reducing the calibration time and the drift influences over the 1 year long measurement time with a SSL approach. The proposed cooperative training algorithm is based on the application of two (or more) neural network regressors, each trained with automatic Bayesian regularization. Differently, the COREG algorithm is based on two K-NN regressors. The cooperating regressors train cooperatively each other by using unlabeled samples. Similarly to the COREG algorithm, at any successive training cycles, each neural network knowledge is updated by extending its training set with samples whose pseudo-labels are predicted by the other cooperating regressor. The samples under evaluation are selected, at each training cycle, from the unlabeled set reservoir. Only the ones, whose inclusion in regressors training set lowers the empirical error computed over its K-nearest labeled neighbours, are actually selected for the inclusion in the other regressor training set. For the two cooperating and equally designed neural networks, we have chosen two different distant metrics for similarity based selection. The first one is based on classic L2 norm while the other is based on grade 5 Minkowski distance. Similarly the K parameter was respectively 3 and 5. Parameters like training duration (expressed in cooperative training cycles) have been selected empirically to 60. In a first experimental setting, we have hidden the presence of drift and other time related issues by randomly extracting training and

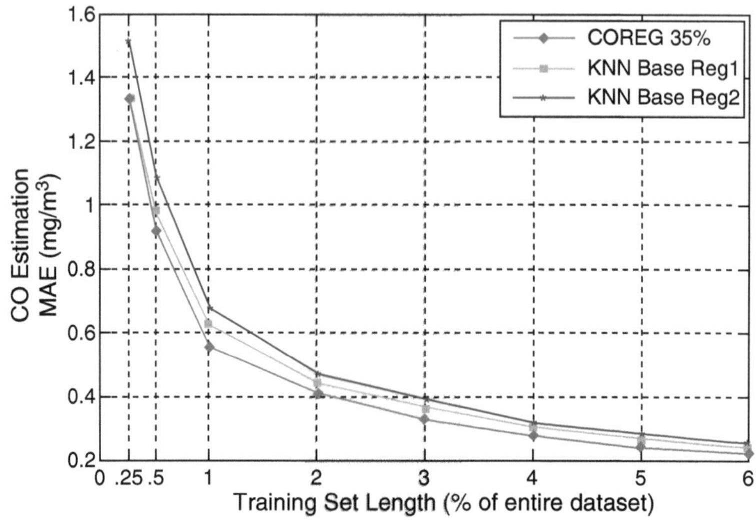

Fig. 27.1 Drift neutral experiment.15 runs averaged CO concentration estimation MAE versus training set length expressed in percent of the dimension of the entire dataset. 35% of the entire database has been used as unlabeled reservoir for standard COREG algorithms. COREG results are compared with both base classifiers results

test samples from the entire dataset. In this way, the drift affected dataset can be treated as a drift neutral problem with noise affected measurements. In this setting we aimed to highlight SSL approach capability to reduce the number of labeled samples needed to build a sufficient knowledge for the multivariate calibration. The MAE (Mean Absolute Error) performance index of the state-of-the-art COREG algorithm has been evaluated on different training and unlabeled sets lengths, comparing it with base regressors performances. Results are shown in Fig. 27.1. COREG algorithm obtains a performance advantage over the two standard K-NN regressors at all the considered training set percentage lengths. It is also worthwhile to note that performance levels obtained by the COREG algorithm at 4% training set length can only be reached by one of the two base regressors at a length of 6% of the entire dataset. The training set length reduction accounts for 139 h of measurements.

In the second experimental setting, we have addressed the performance hit caused by drift issues. We have used the modified cooperative learning algorithm explained above with a filter like methodology to adapt the regressor knowledge over the entire year. The selected supervised training set included only 24 hourly samples (i.e. 1 day, recorded in March), while the unsupervised and test sets included the following U = 400 and T = 24 samples of the entire dataset, respectively (Fig. 27.2).

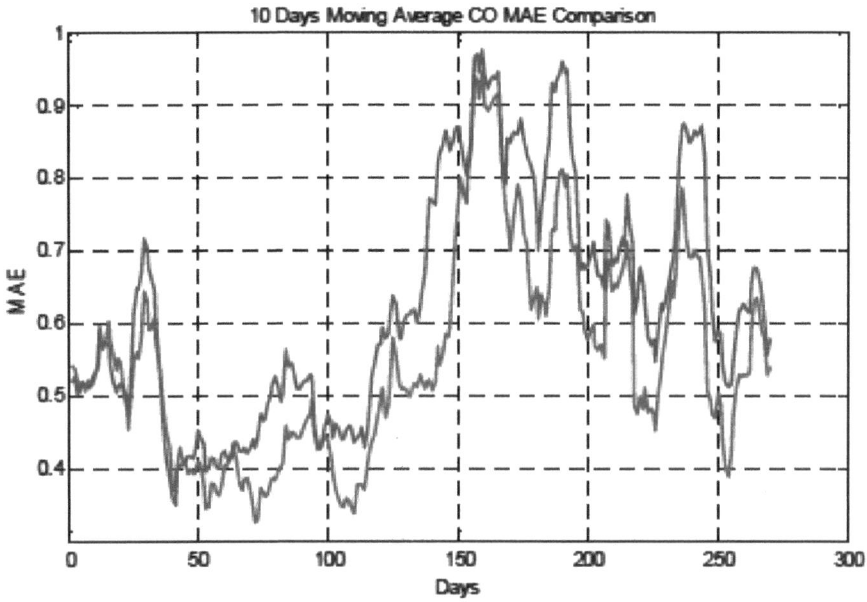

Fig. 27.2 Drift counter action experiment. CO Estimation comparison with integrated KNN-NN SSL algorithm (*red*, 100 unlabeled samples reservoir) and standard NN algorithm (*green*) based on 24 samples. The SSL approach achieved a 14% performance gain with respect to the 1-year long averaged MAE score. The use of moving unlabeled samples has allowed the regressor to adapt for concept and drift effects by using unlabeled samples

After learning stage and performance estimation over the test set, the unsupervised and test sets were T-shifted along all the dataset. Each shift corresponds to a new learning/testing cycle. The MAE performance index is hence computed day-by-day along all the dataset in order to be compared with the one obtained by the best the base neural network architecture trained only with initial 24 samples.

4 Conclusions

We have introduced an original methodology for applying semi-supervised learning in dynamic artificial olfaction problems. We have tested this approach both in a drift neutral and in a drift affected setting by using a large on-field pollution monitoring dataset recorded in a harsh environment. Results show that the proposed approach outperforms the basic supervised approach, commonly applied from the artificial olfaction practitioners in both settings. The cooperative semi-supervised based approach benefitted from the use of updated unlabeled samples, adapting its knowledge to the slowly changing sensor drift effects. By these results, it is reasonable to expect that semi-supervised learning can provide significant advantages to the performance of sensor fusion subsystems in artificial olfaction problems.

References

1. Dockery D, Pope CA, Xu X, Speizer F, Schwartz J (1993) An association between air pollution and mortality in six US cities. N Engl J Med 329:1753–1759
2. Mazzeo NA, Venegas LE (2005) Evaluation of turbulence from traffic using experimental data obtained in a street canyon. Int J Environ Pollut 25:164–176
3. Pearce TC, Schiffman SS, Nagle HT, Gardner JW, Pearce TC, Schiffman SS, Nagle HT, Gardner JW (eds) (2002) Handbook of machine olfaction: electronic nose technology. Wiley-VCH, Weinheim
4. Carotta MC, Martinelli G, Crema L, Malagu C, Merli M, Ghiotti G, Traversa E (2001) Nanostructured thick-film gas sensors for atmospheric pollutant monitoring: quantitative analysis on field tests. Sensors Actuator B Chem 76:336–342
5. De Vito S, Massera E, Piga M, Martinotto L, Di Francia G (2008) On field calibration of an electronic nose for benzene estimation in an urban pollution monitoring scenario. Sensor Actuator B Chem 129:750–757
6. Chapelle O, Scholkopf B, Zien A (2006) Semi-supervised learning. MIT Press, Cambridge
7. Chawla NV et al (2005) Learning from labeled and unlabeled data: an empirical study across techniques and domains. Journal of AI 23:331–366
8. Zhi-Hua Z (2007) Semisupervised regression with cotraining-style algorithms. IEEE Trans Know Data Eng 19(11):1479–1493
9. De Vito S, Piga M, Martinotto L, Di Francia G (2009) CO, NO2 and NOx urban pollution monitoring with on-field calibrated electronic nose by automatic bayesian regularization. SensorActuator B 143:182–191

Chapter 28
Developing Artificial Olfaction Techniques for Contamination Detection on Aircraft CFRP Surfaces: The Encomb Project

Saverio De Vito, Ettore Massera, Grazia Fattoruso, Maria Lucia Miglietta, and Girolamo Di Francia

Composite materials are already used in the manufacturing of structural components in aeronautics industry. However, the light-weight design of Carbon Fiber Reinforced Polymer (CFRP) primary structures is still limited because of the lack of adequate quality assurance procedures for the realisation of the adhesive bonding, which is the optimum technique for joining CFRP light-weight structures. Hence, the primary objective of ENCOMB is the identification, development, and adaptation of methods suitable for the assessment of adhesive bond quality. The performance of adhesive bonds depends on the physicochemical properties of both adherent surfaces and adhesives. Therefore, a set of advanced non-destructive testing techniques is applied and adapted to the characterization of CFRP bonded structures, the state of adherent surfaces before bonding and the state of the cured and uncured adhesives. Actually, surface contamination by several aeronautics fluids eventually results in weak or kissing bonds. The goal of our research work is to investigate solid state chemical sensors and artificial olfaction techniques (AO) for the detection of CFRP surface contamination by aeronautic fluids. The successful implementation of a reliable quality assurance concept within manufacturing and in-service environments will provide the basis for increased use of light-weight composite materials for highly integrated aircraft structures thus minimizing rivet-based assembly. Herein, we present a first approach on the contamination detection scenario, based on the use of an array of polymer sensors.

S. De Vito (✉) • E. Massera • G. Fattoruso • M.L. Miglietta • G. Di Francia
Basic Materials and Devices Department, ENEA – National Agency for New Technologies, Energy and Sustainable Economic Development, Portici (NA), Italy
e-mail: saverio.devito@portici.enea.it

A. D'Amico et al. (eds.), *Sensors and Microsystems: AISEM 2011 Proceedings*, Lecture Notes in Electrical Engineering 109, DOI 10.1007/978-1-4614-0935-9_28,

1 Introduction

During the last decade, Aerospace industry has experimented a growing trend in the use of Carbon Fiber Reinforced Polymer (CFRP) for secondary structure fabrication. CFRP structures can be easily bonded via adhesive assembly procedures. However, the lack of standardized procedures for CFRP adhesive bonds quality assessment has prevented their use in the rivetless assembly of aircraft main structures. Such rivetless assembly could easily lead to extremely lightweight aircraft with very significant savings in fuel usage and CO_2 emissions for commercial air traffic. The development of the so called "green aircraft" is one of the main drivers of FP7 aerospace research programs and of the CleanSky JTI, actually the financially most significant EU research initiative (800MEUR financed).

Within this framework the FP7-ENCOMB (Extensive Non destructive test for Composite Bonds quality assessment) project aims to investigate and develop novel Non destructive test technologies (NDT) for the assessment of both the CFRP surface (pre-bond) and adhesive bonds quality.

The ENCOMB project, lead by the Fraunhofer Institute brings together 14 of the top quality EU research institutions from 6 EU states among which is ENEA, the Italian Agency for New Technologies, Energy and sustainable economic Development. ENEA will focus its activities in investigating and developing artificial olfaction technologies for the contamination detection of adherends surfaces. Actually, contamination, modifying the CFRP wetting behavior, is the main source for weak or kissing bonds that may hamper the bond robustness over time [1]. No specific NDT techniques, apart from the wetting test whose results requires highly skilled personnel [2], have been developed for assessing surfaces contamination. Aeronautics fluids such as hydraulic fluids, water, fuels, release agents, de-icing fluids etc. are the most common contaminants to be considered in this context.

ENEA UTTP-MDB, based in Portici Research Centre, will investigate and develop solid state chemical sensor based electronic nose technologies for the detection, discrimination and quantification of aeronautic fluids contamination on CFRP surfaces.

Selection and development of sampling, sensing and data processing technologies will represent the main effort of the ENEA UTTP-MDB group in the ENCOMB project. The sampling subsystem should be developed to maximize the uptake of volatile molecules from the CFRP surfaces. On the basis of the results of the first experimentation currently carried out with the use of a general purpose electronic nose, the architecture of an ad-hoc sampling system will be designed. Sensors selection will be carried out by investigating different technologies so it will probably lead to a hybrid sensor array; MOX, Polymer, EC, PID and IMS sensing technologies will be screened for their capability to detect main volatile compounds released by contamination agents at significantly low concentration levels (hundreds of ppb). Finally, ad-hoc sensor data processing techniques will be designed and developed to address the different performance of the selected sensor technologies with the aim to enhance discrimination capability and sensor drift resiliency.

In the following, we present the results of a preliminary exploration experiment, useful for the evaluation of the aeronautic fluids discrimination and detection capability of an array of polymer sensors.

2 Results and Discussion

A first Artificial Olfaction scenario was built starting from a polymer sensors array. The selected commercial AO system was based on six, room temperature operating, chemiresistive, polymer sensors array. Sensors were produced by in-situ electropolymerisation of monomers. The sampling subsystem was set in dynamic headspace analysis mode using purified ambient air as carrier gas while volatiles were extracted by bubbling the carrier gas in the sample. Preliminary samples were prepared by making into water emulsion Diesel oil, and an advanced fire resistant aviation hydraulic fluid (0.6% v/v). Output sensors were recorded during a test cycle of 25 s in which the sensors were exposed for 8 s to the vapors analyte. Figure 28.1 shows the multivariate sensors response for the three analytes.

Repeated test cycle with the three analytes were performed and ad-hoc signal features have been extracted, building a suitable dataset for data analysis purposes. Linear Discrimination Analysis (LDA) has been chosen as pattern recognition algorithm in order to create a supervised discrimination map shown in Fig. 28.2. Preliminary results show that the system, after the training, has been capable to detect and discriminate water contamination by hydraulic fluid and Diesel Oil in few seconds.

As matter of fact this is a promising starting point, work in progress deals with sensors stability, training set reliability and estimation of the lowest contamination level detectable by the sensors array.

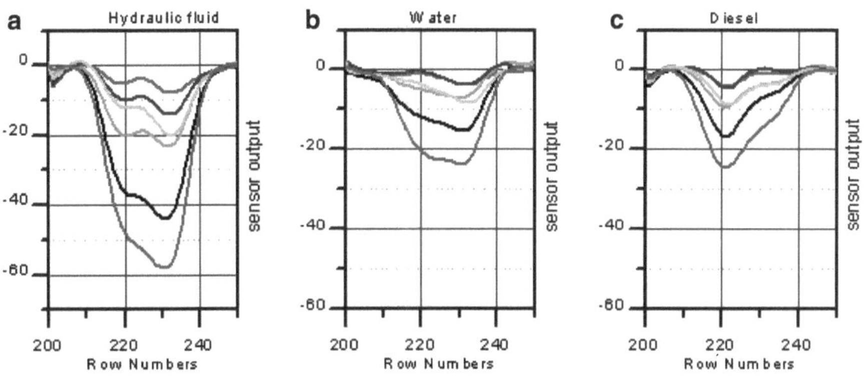

Fig. 28.1 Multivariate sensors outputs during an headspace analysis for (**a**) Hydraulic Fluid emulsified in water. (**b**) Water. (**c**) Diesel oil emulsified in water

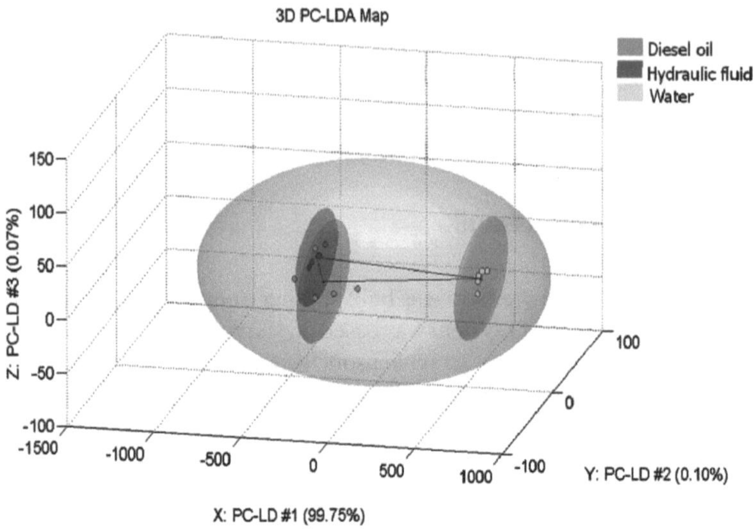

Fig. 28.2 3D training set discrimination map obtained by feature extraction of multivariate sensors output

3 Conclusions

Our preliminary research work on the investigation in solid state chemical sensors and artificial olfaction techniques (AO) for the detection of CFRP surface contamination by aeronautic fluids, produce first experimental results on the capability of a polymer sensors array to detect and discriminate hydraulic fluid contamination in water.

Acknowledgments The research work is funded by EU within the framework of the project ENCOMB (grant no. 266226).

References

1. Davis GD (1993) Contamination of surfaces: origin, detection and effect on adhesion. Surf Interf Anal 20:368–372
2. Markus S, Wilken R, Dieckhoff S, Hennemann O (2006) Quality Monitoring of CFRP Surfaces in Bonding and Coating Processes. A3TS Conference, Bordeaux, 4–6 Oct 2006

Part III
Physical Sensors

Chapter 29
Piezoelectric Polymer Films for Tactile Sensors

Lucia Seminara, Maurizio Valle, Marco Capurro, Paolo Cirillo, and Giorgio Cannata

Some results related to the electro-mechanical characterization in frequency and temperature of PVDF piezoelectric films are reported. A rheological model of creep and recovery of the response of these samples following to a step in load or temperature is suggested. This work is intended as the first step for the electro-mechanical design of innovative integrated transduction systems for tactile sensors in robotic applications.

1 Introduction

Tactile sensing enables robots to perform safe interactions with the environment in case of both voluntary and reactive interaction tasks. In particular the present research concerns sensing technologies and methods for the development of distributed and modular components for general-purpose large-area tactile sensors.[1]

[1] ROBOSKIN European Project, http://www.roboskin.eu

L. Seminara (✉)
Department of Biophysical and Electronic Engineering, University of Genoa, Genoa, Italy
e-mail: lucia.seminara@unige.it

M. Valle
Department of Biophysical and Electronic Engineering, University of Genoa, Genoa, Italy

Research Center on Materials Science and Technology, University of Genoa, Genoa, Italy

M. Capurro • P. Cirillo
Research Center on Materials Science and Technology, University of Genoa, Genoa, Italy

Department of Civil, Environmental and Architectural Engineering
University of Genoa, Genoa, Italy

G. Cannata
Department of Communication Computer and System Sciences, University of Genoa, Genoa, Italy

A. D'Amico et al. (eds.), *Sensors and Microsystems: AISEM 2011 Proceedings*, Lecture Notes in Electrical Engineering 109, DOI 10.1007/978-1-4614-0935-9_29,

This application demands mechanical flexibility, conformability joint with a relatively wide frequency bandwidth ($0 \div 1$ *kHz*).

Piezoelectric transducers in the form of thin polymer films have been chosen [1,2] as they meet these requirements except from perceiving static mechanical stimuli.

The aim is thus to build multisensory systems which integrate different physical sensors on a same patch. In particular our skin system is formed by conformable patches of triangular shape, interconnected in order to form a networked structure [3]. The piezoelectric "functional" material must be integrated into complex mechanical structures which also include a substrate and a protective layer. How to integrate the PVDF transducer is not an easy task, because its behavior depends on several aspects including the properties of the whole mechanical chain, such as, e.g., material and thickness of the protective layer. Moreover, these design features also influence the requirements of the interface electronics and the data processing, to cite some of the most important aspects. In the present work we start from an understanding of the transducer behavior in different basic conditions in order to orient the design of complex systems building the robotic skin.

Piezoelectric materials intrinsically convert the mechanical stimulus into an electrical signal on the basis of their electromechanical properties which are expressed by a number of coefficients including elastic, dielectric and piezoelectric moduli. These properties are affected by temperature and frequency. Therefore, in-depth understanding of PVDF dynamic response under a variety of operating conditions is critical in view of achieving the desired performance outcomes. Available data, coming from literature [4–6], as well as from manufacturers, is not sufficient for an effective and reliable assessment. In the present work the frequency behavior of the d_{33} piezoelectric modulus is presented and some considerations are made on the PVDF response time-temperature dependence.

2 Results and Discussion

Commercial PVDF sheets – already stretched and poled – have been purchased from Measurement Specialties Inc.[2] Stretching at temperatures well below the melting point of the polymer causes chain packing of the molecules into parallel crystal planes ("beta phase"). The beta phase polymer is poled by application of electric fields of the order of *100* V/μm to align the crystallites to the poling field. In such conditions the piezoelectric behavior exhibits a material symmetry in the orthorhombic crystal system (C2V class), corresponding to that of the so-called orthotropic materials.

[2] http://www.meas-spec.com/default.aspx

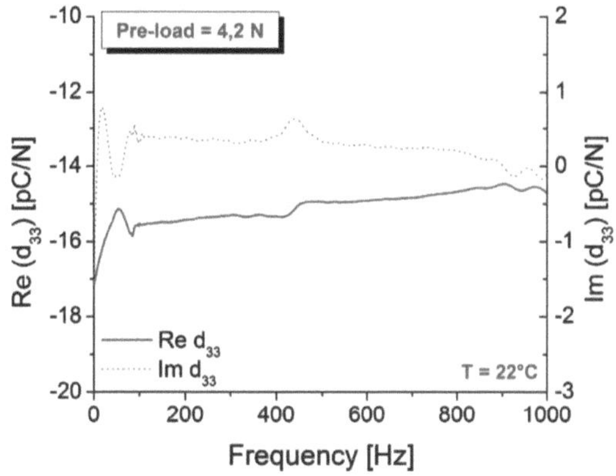

Fig. 29.1 Frequency behavior of the d_{33} piezoelectric modulus

Linear electro-elastic constitutive equations are commonly used to describe the coupling of dielectric, elastic, and piezoelectric properties in piezoelectric materials [7]. The experimental setup and measurement methods to retrieve the complex moduli are extensively described in [8]. When working in thickness mode, as common in most applications, the d_{33} element of the piezoelectric matrix relates the charge density on the film surface to the through-thickness normal stress T_3 [8]. d_{33} measurements have been performed using a shaker to compress the PVDF film in the thickness direction. Conductive glue has been used to create a uniform contact between the sample and the heads of the testing machine, thus excluding possible variations in the contact area during measurements. Tests at different preloads have been performed and results confirmed that observed nonlinearities were due to variations in the contact area and not to intrinsic properties of the sample as a function of the applied stress (Fig. 29.1).

The d_{33} modulus is almost flat in frequency apart of small disturbances due to the experimental apparatus (50 Hz noise and small electric resonances). This modulus is an essential design parameter for robot skin models where surface dynamic stimuli are transmitted through a deformable layer to a sensor grid placed at the bottom.

Not only frequency, but also temperature affects the behavior of the moduli. As an example, we report the temperature dependence of the frequency spectrum of one of the piezoelectric moduli (d_{31}) (Fig. 29.2).

Moreover, once subjected to a step temperature variation, the value of the modulus tends to recover in very long times. Both these aspects are relevant for the present application.

A rheological model of creep and recovery of the response of PVDF samples following to a step in load or temperature has been examined.

Starting from a standard solid model which roughly describes the time response of viscoelastic materials to a loading time step, a more advanced model has been set up to take into account the observation of two different regimes in the sample behavior, corresponding to a "short" and a "long" time responses. The long time relaxation cannot be detected by dynamic measurements, because it is difficult to

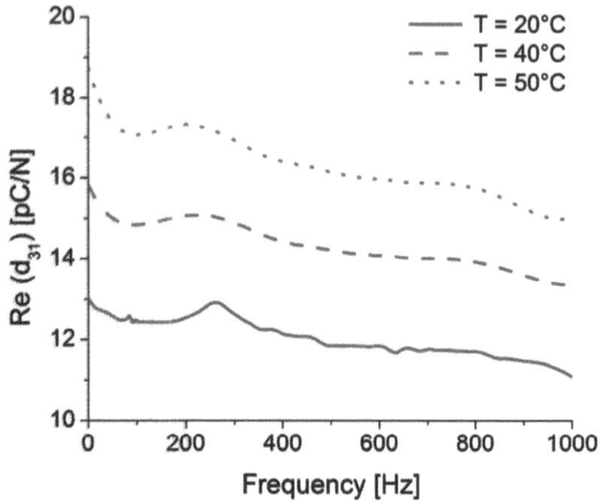

Fig. 29.2 Temperature dependence of the frequency spectrum of the d_{31} piezoelectric modulus

resolve very low frequencies ($\omega << 1$). Conversely, dynamic effects are not detected in conventional creep or relaxation tests. Conceptually, this behavior can be simulated by coupling (e.g. in parallel) two standard solid models, one with a high relaxation time (τ_1, may be of the order of thousands of seconds) and the other with a very short relaxation time (τ_2, say of the order of milliseconds). The same behavior has been extended to the sample response to temperature changes. A more detailed analysis will be contained in a forthcoming publication.

3 Conclusions

The d_{33} piezoelectric modulus which characterizes the electromechanical response of PVDF working in thickness mode has been measured in the $1 \div 1000$ *Hz* frequency range. The temperature dependence of piezoelectric moduli has been discussed. These measurements enforce the integration of a temperature reading on site.

The knowledge established for single films will be exploited both to design suitable embedded electronics at a system level and to optimize the overall behavior of innovative devices based on PVDF sensors. Our approach is based on a structured scheme consisting of a combination of experimental materials and transducers characterization and Multiphysics Finite Element Analysis (FEA) simulations. A series of experiments has been planned on integrated model devices to be validated by FEA simulations.

Acknowledgments This work has been supported by the European Commission project "ROBOSKIN", under grant agreement no. 231500.

References

1. Dahiya RS, Metta G, Valle M, Sandini G (2010) Tactile Sensing: From Humans to Humanoids. IEEE Trans Robot 26:1–20
2. Lee H-K, Chang S-I, Yoon E (2006) A flexible polymer tactile sensor: fabrication and modular expandability for large area deployment. J Microelectromech Syst 15:1681–1686
3. Sandini G, Maggiali M, Cannata G, Metta G (2007) Tactile sensor arrangement and corresponding sensory system. US Patent No. 20100234997
4. Vinogradov AM, Holloway F (1999) Electro-mechanical properties of the piezoelectric polymer PVDF. Ferroelectrics 226:169–181
5. Roh Y, Varadan VV, Varadan VK (2002) Characterization of all the elastic, dielectric and piezoelectric constants of uniaxially oriented poled PVDF films. IEEE Trans Ultrason Ferroelectr Freq Control 49:836–847
6. Kwok KW, Chan HLW, Choy CL (1997) Evaluation of the material parameters of piezoelectric materials by various methods. IEEE Trans Ultrason Ferroelectr Freq Control 44:733–742
7. Ikeda T (1996) Fundamentals of piezoelectricity. Oxford Science, New York, Tokio
8. Seminara L, Capurro M, Cirillo P, Cannata G, Valle M (2011) Electromechanical characterization of piezoelectric PVDF polymer films for tactile sensors in robotics applications. Sensor Actuator A 169:49–58

Chapter 30
An Ultra High Sensitive Current Sensor Based on Superconducting Quantum Interference Device

A. Vettoliere, C. Granata, B. Ruggiero, and M. Russo

The developed sensor design is based on a suitable superconducting intermediary (matching) magnetic flux transformer magnetically coupled to a niobium based dc-SQUID (Superconducting Quantum Interference Device). The 60 square niobium turns (20 μm width) signal coil is tightly coupled to the matching transformer consisting of a square single turn primary coil connected in series with a multiturn secondary coil. The obtained signal current to magnetic flux transfer factor (current sensitivity) is equal to 62 nA/Φ_0 measured by using a current sensing noise thermometer technique. The sensor has been characterized in liquid helium by using a direct coupling low noise readout electronic and the flux locked loop configuration. Despite the circuit complexity, the sensor has exhibited a smooth and free resonance voltage-flux characteristic ensuring a stable working operation. Considering a SQUID magnetic flux noise $\sqrt{S_\Phi} = 1.8\ \mu\Phi_0/\sqrt{Hz}$ at T = 4.2 K, a current noise as low as 110 fA/√Hz is obtained. Due to his high performance such sensor can be employed in all application requiring an extremely current sensitivity, like the readout of the gravitational wave detectors and the current sensing noise thermometry.

1 Introduction

Due to their extreme magnetic field sensitivity, Superconducting QUantum Interference Devices (SQUID) are widely used in several applications [1] like biomagnetism, magnetic microscopy, non-destructive evaluation, astrophysics and, recently, also in nanoscience [2]. These sensors are also employed to detect very small changes in various physical quantities which can be transformed into changes in the magnetic flux treading the SQUID ring. Among them, particular

A. Vettoliere • C. Granata (✉) • B. Ruggiero • M. Russo
Istituto di Cibernetica "E. Caianiello" del Consiglio Nazionale
delle Ricerche, Pozzuoli (Napoli), Italy
e-mail: c.granata@cib.na.cnr.it

A. D'Amico et al. (eds.), *Sensors and Microsystems: AISEM 2011 Proceedings*, Lecture Notes in Electrical Engineering 109, DOI 10.1007/978-1-4614-0935-9_30,

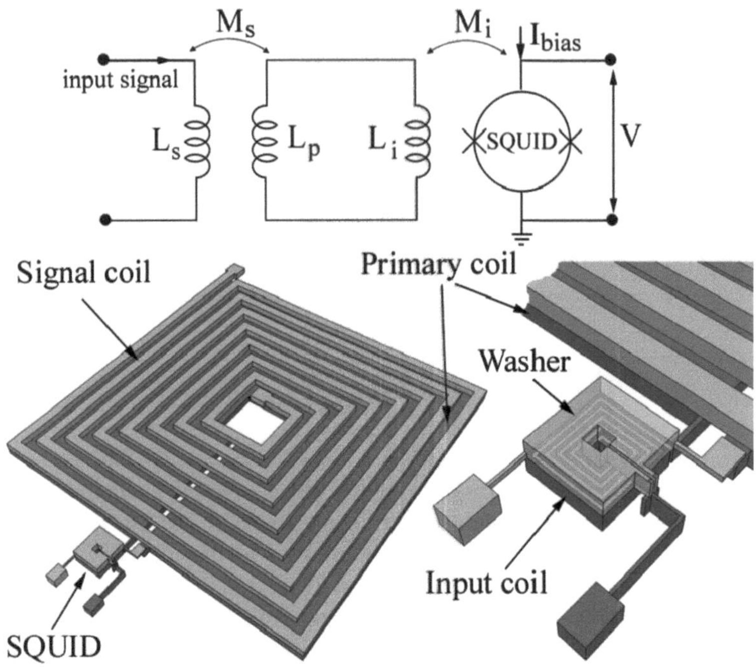

Fig. 30.1 Electrical scheme of a SQUID current sensor based on a double transformer (*upper figure*). Sketch of the device including the signal coil, the intermediary flux transformer and the SQUID in a washer configuration (*lower figure*)

interest is devoted to the SQUID based current sensors because they can be effectively employed as readout of gravitational wave detectors [3], transition-edge sensors [4] and for current sensing noise thermometry [5]. The best way to obtain a practical and reliable SQUID current noise is the design based on a double transformer coupling [6, 7] consisting in a matching (intermediary) transformer inserted between the signal coil and the input terminals of the conventional configuration. Here an ultra high sensitive SQUID current sensor based on an optimized double transformer is presented. With respect to single transformer it allows to efficiently couple the low-inductance SQUID with a very high signal coil inductance (tenths of mH), obtaining an ultra low electric current noise.

1.1 Sensor Design and Fabrication Process

The fully integrated device has an area of 1 cm^2 and it includes a signal coil, an intermediary flux transformer, a SQUID in a washer configuration, a feedback coil for Flux Locked Loop (FLL) operation and a thin film resistor network useful for the current noise characterization of the sensor. The signal coil, consisting of a superconducting 60-turns coil, is magnetically coupled to the intermediary transformer (Fig. 30.1). The primary coil of the intermediary

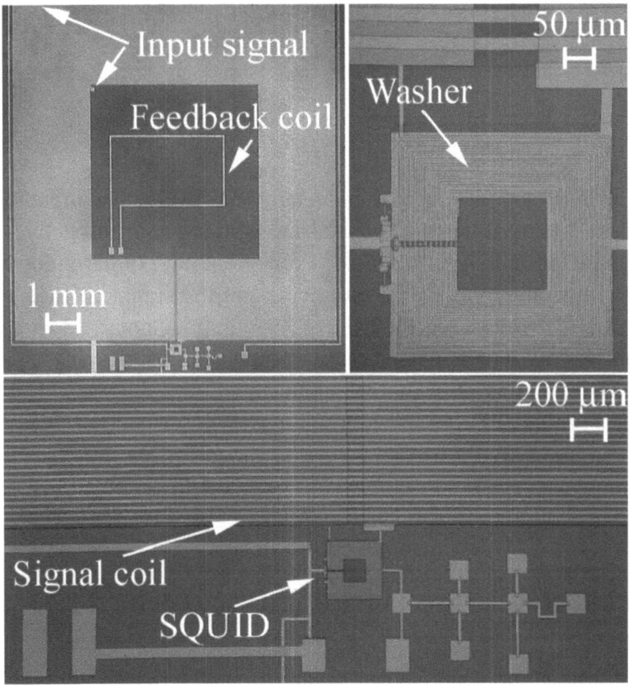

Fig. 30.2 Fully integrated SQUID current sensor with a double transformer. The sensor includes a 60 turn-signal coil a superconducting intermediary flux transformer, the SQUID in a washer configuration and a feedback coil for Flux Locked Loop operations

transformer consists of a single superconducting square coil having a side length of 10 mm and a large side width (2.5 mm) in order to accommodate the signal coil having a large turn's number. The primary coil ($L_p = 9.3$ nH) is connected in series with a 12-turn input coil ($L_i = 33$ nH), which is coupled to the SQUID loop in a washer configuration having a large inductance L = 250 pH, in order to increase the flux gain [8]; the consequent performance degradation, due to a non-optimal β_L value (>1) is avoided by inserting a damping resistor across the SQUID inductance [9]. A resistive feedback rectangular coil is located in the primary coil's hole (Fig. 30.2) and acts also as device heater in the case of entrapped flux.

An important figure of merit for a SQUID current sensor is the input current to magnetic flux transfer factor $I_\Phi = 1/M$ (current sensitivity); that is the input current value to send in the signal coil to couple a flux quantum ($\Phi_0 = 2.07 \times 10^{-15}$ Wb) in the SQUID loop. The spectral density of the current noise is related to the SQUID magnetic flux noise by a simple expression: $S_I^{\ 1/2} = I_\Phi \cdot S_\Phi^{\ 1/2}$. The current sensitivity can be easily calculated by evaluating the current to send in the signal coil to generate a magnetic flux in the primary coil that matches a flux quantum in the SQUID:

$$i_s = \frac{(L_p + L_i)\,\Phi_0}{M_i M_s} = \frac{(L_p + L_i)\,\Phi_0}{k_i\,\sqrt{LL_i}\,k_s\,\sqrt{L_s L_p}} \Rightarrow I_\Phi = \frac{i_s}{\Phi_0}$$

$$\cong \frac{(L_p + L_i)}{(k_i \cdot k_s \cdot n_i \cdot n_s \cdot L \cdot L_p)} \tag{30.1}$$

Where M_i and M_s are the mutual inductances between the input coil and the SQUID and between the signal coil and the primary one respectively, L_s is the signal coil inductance, k_i and k_s are the corresponding coupling constants and n_i and n_s are the turn numbers of the input coil and the signal coil respectively. By using the numerical values reported above for the inductances and a value of 0.95 for both coupling constants, we can estimate a current responsivity value as low as 60 nA/Φ_0 corresponding to a mutual inductance M = 34.5 nH. The fabrication process, based on the niobium technology, is well described in ref. 8. A picture of the fully integrated SQUID current sensor is shown in Fig. 30.2.

2 Experimental Performance and Discussion

The SQUID sensor has been characterized in liquid helium at T = 4.2 K in a coaxial double shield (lead and cryoperm) using a direct coupling low noise read-out electronic. The voltage-magnetic flux characteristic (V-Φ), the input current-magnetic flux transfer factor (I_Φ) and the spectral density of magnetic flux noise ($\sqrt{S_\Phi}$) has been measured. The critical current and the normal resistance of the SQUID sensor are respectively $2I_c = 24\ \mu A$ and $R = R_s/2 = 1.8\ \Omega$ corresponding to a $\beta_C = 2\pi I_c C R_s^2/\Phi_0$ value of 0.8 (C = 1.7 pF is the capacitance of a 20 μm^2 Josephson junction). In Fig. 30.3, V-Φ characteristic showing a large voltage swing (ΔV = 60 μV) is reported. The maximum responsivity measured on the steeper side of the characteristic is $V_\Phi = 310\ \mu V/\Phi_0$.

The measurement of the current sensitivity I_Φ is based on a current sensing noise thermometer technique. Closing the signal coil on the integrated test

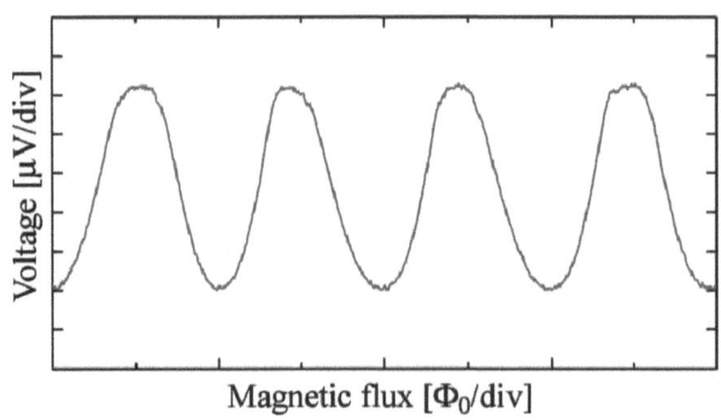

Fig. 30.3 Voltage-flux characteristic (V-Φ) of the current sensor measured at T = 4.2 K

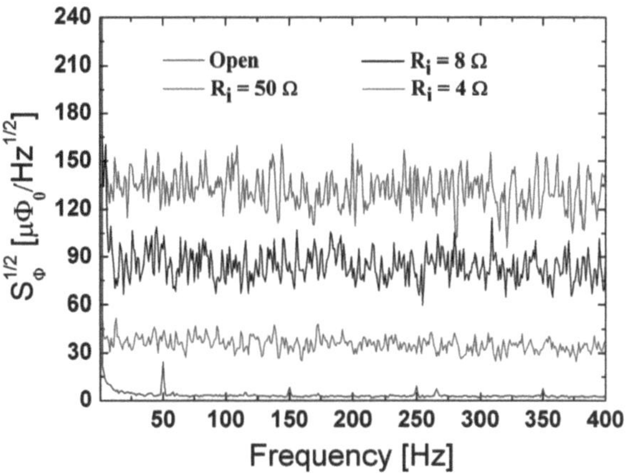

Fig. 30.4 Sensor magnetic flux spectra measured at T = 4.2 K in flux locked loop configuration, using a direct coupling scheme with a low noise readout electronics. The lower curve refers to the sensor noise when the signal coil is open. The other curves correspond to the Nyquist noises induced by three different test resistors (50, 8 and 4 Ω) connected to the signal coil

resistors, the current generated by a Nyquist noise induces a magnetic flux noise in the SQUID:

$$\sqrt{S_\Phi} = \frac{1}{I_\Phi}\sqrt{\frac{4k_BT}{R_t}} \Rightarrow I_\Phi = \sqrt{\frac{4k_BT}{S_\Phi R_t}} \qquad (30.2)$$

Where R_t is the resistance value and k_B is the Boltzman constant. In Fig. 30.4, the spectra of the magnetic flux noise measured at T = 4.2 K in FLL configuration relative to different values of the test resistors are reported. The lower spectrum, relative to open signal coil pads, corresponds to the sensor noise (2.9 $\mu\Phi_0/\sqrt{\text{Hz}}$) while the intrinsic one obtained subtracting the amplifier contribution, is 1.8 $\mu\Phi_0/\sqrt{\text{Hz}}$. By entering in Eq. (30.2) the sensor noise spectral densities relative to different resistor values, a value of $I_\Phi = 62$ nA/Φ_0 is obtained, in excellent agreement with the prediction of Eq. (30.1). Considering the SQUID intrinsic flux noise, a current noise spectral density of $\sqrt{S_I} = 110$ fA/$\sqrt{\text{Hz}}$ is obtained which is about three time smaller than the noise of other SQUID of the same category [10].

3 Conclusions

The sensor design, based on an optimized double transformer, has been finalized to get both a very low current to magnetic flux transfer factor and a suitable voltage to flux characteristic. The effectiveness of the device, confirmed by the experimental results, makes it suitable in all applications based on the measurement of ultra small electric currents.

References

1. Blake RC, Wellstood FC (2006) in The SQUID handbook vol II: applications of SQUIDs and SQUID systems. Clarke J, Braginski AI(eds) Wiley-VCH Verlag GmbH & Co. KgaA, Weinheim, p 391; Fagaly RK (2006). Superconducting quantum interference device instruments and applications. Rev Sci Instrum 77:101101
2. Foley CP, Hilgenkamp H (2009) Why NanoSQUIDs are important: an introduction to the focus issue. Superc Sci Technol 22:064001
3. De Waard A, Gottardi L, Van Houwelingen J, Shumack A, Frossati G (2003) MiniGRAIL, the first spherical detector. Class Quantum Grav 20:143
4. Chervenak JA, Irwin KD, Grossman EN, Martinis JM, Reintsema CD, and Huber ME (1999) Superconducting multiplexer for arrays of transition edge sensors. Appl Phys Lett74:4043; Jongsoo Y, Clarke J, Gildemeister JM, Lee AT, Myers MJ, Richards PL, and Skidmore JT (2001). Single superconducting quantum interference device multiplexer for arrays of low-temperature sensors. ibid78:371; Irwin KD and Lehnert KW(2004). Microwave SQUID multiplexer. ibid85:2107
5. Webb RA, Giffard RP, Wheatley JC (1973) Noise Thermometry at Ultralow Temperatures. J Low Temp Phys 13:383; Lusher CP, Junyun Li, Maidanov VA, Digby ME, Dyball H, Casey A, Ny´eki J, Dmitriev VV, Cowan BP, Saunders J (2001). Current sensing noise thermometry using a low Tc DC SQUID preamplifier. Meas Sci Technol12:1
6. Muhlfelder B, Johnson W, Cromar MW (1983) Double transformer coupling to a very low noise SQUID. IEEE Trans Magn 19:303
7. Polushkin V, Gu E, Glowacka D, Goldie D, Lumley J (2002) A tightly coupled dc SQUID with an intermediary transformer. Physica C 367:280
8. Granata C, Vettoliere A, Russo M (2007) Miniaturized superconducting quantum interference magnetometers for high sensitivity applications. Appl Phys Lett 91:122509
9. Enpuku K, Muta T, Yoshida K, Ire F (1985) Noise characteristics of a dc SQUID with a resistively shunted inductance. J Appl Phys 58:1916; Enpuku K, Yoshida K, Kohjiro S (1986) Noise characteristics of a dc SQUID with a resistively shunted inductance. II. Optimum damping. J Appl Phys 60:4218
10. Pleikies J, Usenko O, Frossati G, Flokstra J (2009) Optimization of a low-Tc dc SQUID amplifier with tightly coupled input coils. IEEE Trans Appl Supercond 19:199

Chapter 31
Tactile Sensing Systems Based on POSFET Sensing Arrays

R.S. Dahiya, D. Cattin, A. Adami, C. Collini, L. Barboni, M. Valle, L. Lorenzelli, R. Oboe, G. Metta, and F. Brunetti

This work presents the current achievements on the POSFET sensing arrays based tactile sensing system. The tactile sensing chips implement POSFET (Piezoelectric Oxide Semiconductor Field Effect Transistor) devices arrays and temperature sensors. This work presents quantitative evaluation of the tactile sensing chip and a proposal for the electronic data acquisition system. Our goal is to integrate the POSFET arrays into the hands of humanoid robots.

1 Introduction

Future robots will work closely and interact safely by using the sense of touch. While it is desirable to have tactile sensors over whole body, the robotic hands are accorded higher priority due to their involvement in majority of the daily tasks.

R.S. Dahiya • A. Adami • C. Collini • L. Lorenzelli
Bio-MEMS, FBK, Trento, Italy
e-mail: dahiya@fbk.eu; andadami@fbk.eui; ccollini@fbk.eu; lorenzel@fbk.eu

D. Cattin • R. Oboe
Department of Management and Engineering, University of Padova, Vicenza, Italy
e-mail: dvd.c@unipd.it; roberto.oboe@unipd.it

L. Barboni • M. Valle (✉)
Department of Biophysical and Electronic Engineering, University of Genova
UNIGE, Genoa, Italy
e-mail: l.barboni@unige.it; maurizio.valle@unige.it

G. Metta
RBCS, Italian Institute of Technology, Genova, Italy

F. Brunetti
Department of Electronic Engineering, Engineering University of Rome Tor Vergata, Rome, Italy
e-mail: francesca.brunetti@uniroma2.it

A. D'Amico et al. (eds.), *Sensors and Microsystems: AISEM 2011 Proceedings*,
Lecture Notes in Electrical Engineering 109, DOI 10.1007/978-1-4614-0935-9_31,

Over the years, tactile sensing technology has improved and many force/pressure sensors and sensing arrays using various materials and transduction methods have been developed [1, 2]. We have designed and manufactured arrays of 1 × 1 mm sized POSFET (Piezoelectric Oxide Semiconductor Field Effect Transistor) devices [1, 3]. The work presented here is an assessment of current research results on effectively integrating POSFET arrays into the robot mechanical and electronic system. This contribution further extends our research on POSFET touch sensing devices [4, 5] towards the design and implementation of tactile sensing system on chip.

2 Experimental Results

The POSFET tactile sensor has been designed to mimic the human fingertips sense of touch [5]. The fabricated sensor chip consists of an array of 5×5 sensing elements, called taxels. To quantify the gain/sensitivity and identify the tactile sensor, an experimental setup has been designed and tests in time and frequency domain have been performed. The sensor has been tested applying a normal force with different patterns (sinusoidal, square and triangular wave) and there was almost no delay between input and output [6], showing a quick response. To test the sensitivity, a sinusoidal normal force at 75 Hz with variable amplitude has been applied. The sensor has been tested with and without the protective layer of PDMS, used to protect it from wear and tear, and in both cases the sensor showed good linearity, as reported in Fig. 31.1.

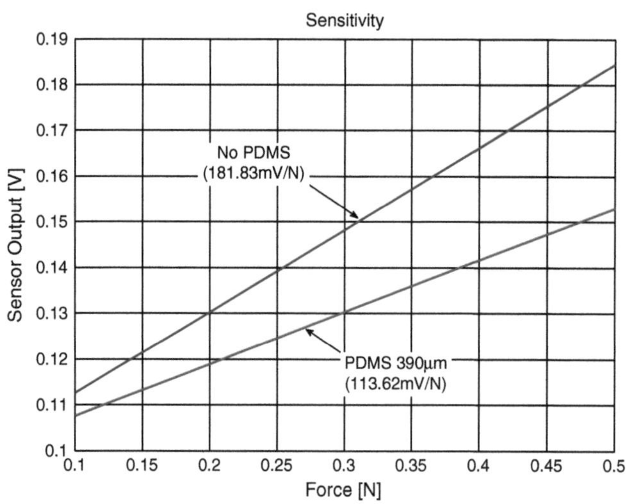

Fig. 31.1 Sensitivity of POSFET touch sensors without PDMS cover and with 390 μm thick PDMS cover

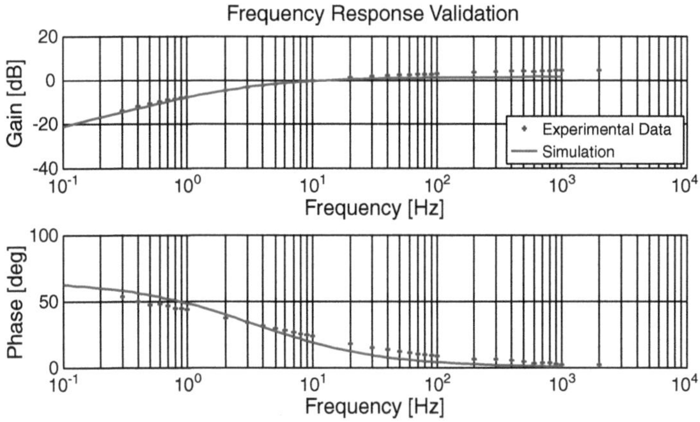

Fig. 31.2 The experimental and simulated frequency response

The identification process has been based on the frequency response which has been measured applying a pre-load of 1 N and a normal sinusoidal force of constant amplitude (0.1 N) at variable frequency from 0.2 Hz to 2 kHz. This range of frequency is wide enough to identify the sensor, because it is higher then the maximum frequency perceptible by the human skin, which has a band pass of about 1.5 kHz [7]. The POSFET frequency response, reported in Fig. 31.2, shows a nonlinear behavior at low frequency, where the slope is almost 10 dB/dec instead of the usual multiple of 20 dB/dec. This behavior has been ascribed to the piezoelectric polymer due to phenomena of mechanical viscoelasticity and dielectric relaxation, which are well described using fractional order systems [8, 9]. Using a modified Cole-Cole function, it is possible to describe the nonlinear behavior of the piezoelectric polymer [10].

A number of measurements were also made by applying the force simultaneously on multiple POSFETs tactile devices. While force was applied on selected POSFETs, the output of all POSFETs on the chip was recorded. The probes of dimension 1 $\times$ 1 mm were made with Eden250 3D printing system, which provides high quality rapid prototyping with typical tolerance of 100 μm. Measuring the response of all POSFETs, while force is applied on selected few, gives a measure of cross–talk among the sensing elements.

The response of various taxels, when a 670 Hz sinusoidal force was applied on the POSFETs (1,3), (1,4), (1,5), (2,4), and (3,4) that together make a 'T' shape, is shown in Fig. 31.3 (a snap-shot of the bar and binary images, obtained from the normalized response of various taxels, is also shown in Fig. 31.3). The taxels pressed by probe can be differentiated from others. The variation among the responses of the POSFETs that were pressed is low (maximum of 18.1 mV recorded for taxel (3,4) and minimum 16.4 mV from taxel (1,3)) and as expected they are in phase.

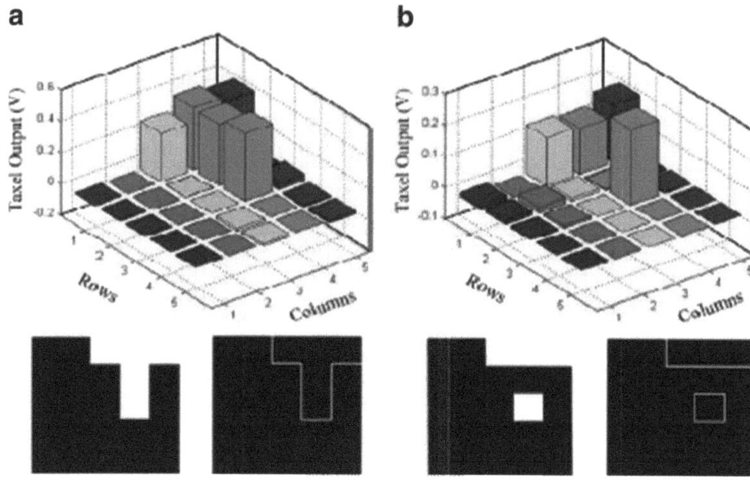

Fig. 31.3 The POSFETs response when is applied: (**a**) 'T' shaped probe. (**b**) Second probe

Similar results were obtained with 20, 120, 370 Hz sinusoidal forces. The same experiment was repeated with the second probe. The snapshot of the bar and binary images, obtained from the normalized response of various taxels, is shown in Fig. 31.3. The negligible response of taxel (2,4), when second probe is used, shows the high spatial resolution (<1 mm) that can be obtained with tactile sensing chips presented here. The data obtained from these experiments was also used to detect the edges by using gradient operators.

3 Tactile Sensing Electronic Data Acquisition System

The designs of read circuitry and test boards are influenced by a number of factors such as biasing configuration of POSFET devices, number of sensing elements on the chip, application requirements, availability of space on robots body, robot communication bus bandwidth, etc.. As an example, the POSFET array must be scanned is such a way that the response of various sensing elements to contact forces having frequency contents up to 1 kHz can be successfully recovered. Further, the POSFET devices must be used in such a way that they respond to the full range of contact forces (0.01–10 N) before MOS transistor switches to a non–operational mode. Keeping in view these issues the data acquisition system has been designed and the block diagram of the system is shown in Fig. 31.4. The data acquisition system, which is to be implemented with commercial off–the–shelf components, consists of three major blocks – clearly marked in the Fig. 31.4.

The block I consists of POSFET tactile sensing array and the components for biasing the POSFET devices.

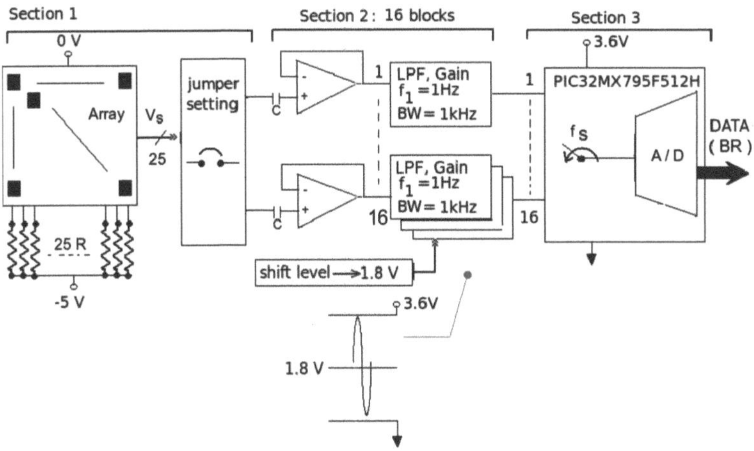

Fig. 31.4 The block diagram of the data acquisition system

The biasing of POSFETs in block I is done using resistors in place of current sources. A bank of R = 20 kΩ resistors is thus used in the block I of Fig. 31.4. The block II is the signal conditioning block. The implemented features are: (a) ac coupling between POSFET outputs and following signal conditioning stages; (b) low pass filtering of the POSFET output signal; (c) very high input impedance; (d) voltage signal range translation. The ac coupling of the POSFET circuit output voltage removes the undesirable effects related to the technology spread among the POSFET devices.

The ac coupling and the low pass filtering together make a band pass frequency filter with the first and second cut–off frequencies placed at 1 Hz and 1 kHz.

This feature helps to define the signal frequency bandwidth and to avoid frequency spectra aliasing during sampling. The input impedance of the signal conditioning block is very high in the signal frequency band of interest (i.e. from 1 Hz to 1 kHz). The block III, in Fig. 31.4, consists of microcontroller PIC32MX795F512H from Microchip Inc. The key microcontroller features are: up to 16 analog inputs, operating voltage range 2.3–3.6 V and 10 bits AD converter (resolution of 3.5 mV @ 3.6 V full scale voltage reference). Since this microcontroller has up to 16 analog input pins, 16 POSFETs on the array can be used with this data acquisition board. The main parameters of the data acquisition system are (see Fig. 31.4) data output bit rate BR = 1.6 Mbits/s, single taxel sampling frequency f_s =10 kHz.

4 Future Work

A preliminary study on the realization of O-POSFET (Organic POSFET) on a flexible substrate using pentacene as organic active layer has been performed.

In the architecture analyzed, the piezoelectric layer is used both as insulator and as substrate for the device in a bottom-gate, bottom-contact configuration. A variation of the piezoresistive behavior of the devices depending on the polarization of the film has been observed. As a future step the piezoelectric response of the device will be studied opening the way to the realization to a flexible tactile sensor in which both the dynamic and the static contact forces will be measured.

Acknowledgment This work is supported by the European Commission Project ROBOSKIN (Grant No. ICT-FP7-231500)
The research leading to these results has received funding from European Community's 7th framework "People" specific programme – Marie Curie Actions – under grant agreement PCOFUND–GA–2008–226070

References

1. Dahiya RS, Metta G, Valle M, Sandini G (2010) Tactile sensing from humans to humanoids. IEEE Trans Robot 26(1):1–20
2. Cutkosky MR et al (2008) Force and tactile sensors. In: Siciliano B, Khatib O (eds) Springer handbook of robotics. Springer, Berlin Heidelberg, pp 455–476
3. Dahiya RS et al (2009) Piezoelectric oxide semiconductor field effect transistor touch sensing devices. Appl Phys Lett 95: 0340105 (1–3)
4. Dahiya RS et al (2009) Design and fabrication of POSFET devices for tactile sensing, in Transducer 2009. The 15th IEEE international conference on solid-state sensors, actuators and microsystems, Denver, pp 1881–1884
5. Dahiya RS, Valle M, Oboe R, Cattin D (2009) Development and characterization of touch sensing devices for robotic applications, IEEE-IECON 2009. The 35th annual conference of the IEEE industrial electronics society, Porto, 3–5 Nov 2009, pp 4245–4250
6. Dahiya RS (2008) Touch sensor for active exploration and visuo-haptic integration. PhD thesis dissertation, Genova
7. Liu JG, Xu MY (2006) Higher-order fractional constitutive equations of viscoelastic materials involving three different parameters and their relaxation and creep functions, mech. Time-Depend Mater 10:263–279
8. Schiessel H, Metzler R, Blumen A, Nonnenmacher TF (1995) Generalized viscoelastic models: their fractional equations with solutions. J Phys A: Math Gen 28:6567–6584
9. Cattin D, Oboe R, Dahiya RS, Valle M (2010) Identification and validation of a lumped parameters model for the dielectric relaxation of a piezoelectric tactile sensor IEEE-ISIE2010. International symposium on industrial electronics, Bari, , 4–7 July 2010, pp 452–457
10. Heywang W, Wersing W, Lubitz K (2008) Piezoelectricity evolution and future of a technology, Springer series in materials science. Vol 114, Springer-Verlag, Berlin Heidelberg. Springer, New york

Chapter 32
POSFET Touch Sensing Devices: Bias Circuit Design Based on the ACM MOS Transistor Compact Model

L. Barboni, M. Valle, and R.S. Dahiya

In this article we present a common-drain floating gate bias circuit design for POSFET (Piezoelectric Oxide Semiconductor Field Effect Transistor) touch sensing devices. A graphical-aided methodology, intended to furnish a criterion that assures the selection of the most appropriate bias resistance value, is presented.

The methodology utilizes the Advanced Compact MOSFET (ACM) model and the g_m characteristic as function of I_D. A design space map is generated showing the trade-offs between gain, current consumption and components value (i.e. resistance). As a proof of concept of the proposed method, a design example and measurements are presented.

1 Introduction

This work deals with POSFET tactile sensing chip and the POSFET bias circuit scheme for the read out circuit design. The read out circuitry is needed to acquire the tactile data from the POSFET tactile sensing chip that is to be integrated on the robot's hands.

The POSEFT tactile sensing chip, a part of which is shown in Fig. 32.1, comprises of a 5 x 5 array of POSFET touch sensing devices. Each POSFET taxel (Touch sensing element) on the chip is 1 mm^2 in size. POSFET devices are obtained by spin coating piezoelectric polymer P (VDF-TrFE), poly (vinylidene fluoride-trifluoroethylene), film on the gate area of MOS (Metal Oxide Semiconductor) transistor.

L. Barboni • M. Valle (✉)
Department of Biophysical and Electronic Engineering, University of Genova, Genova, Italy
e-mail: l.barboni@unige.it; maurizio.valle@unige.it

R.S. Dahiya
Bio-MEMS, Fondazione Bruno Kessler, Trento, Italy
e-mail: dahiya@fbk.eu

A. D'Amico et al. (eds.), *Sensors and Microsystems: AISEM 2011 Proceedings*, Lecture Notes in Electrical Engineering 109, DOI 10.1007/978-1-4614-0935-9_32,

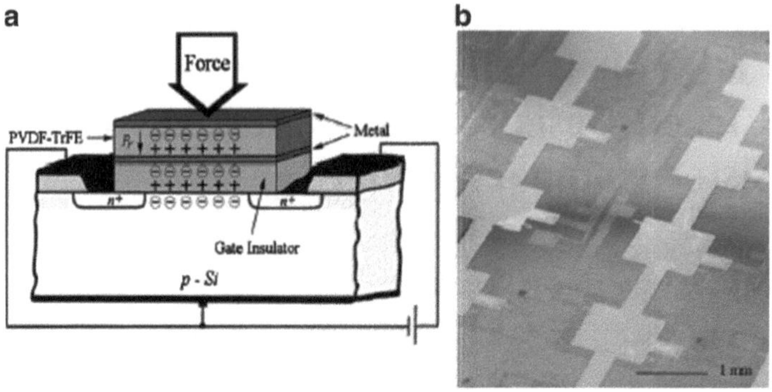

Fig. 32.1 (**a**) The structure of a POSFET touch sensing device. (**b**) A part of 5 x 5 POSFET tactile sensing array, after fabrication. Each POSFET taxel on the array has 1 mm^2 in area [1–3]

To detect contact events, the taxels utilize the contact forces induced change in the polarization level (and hence the change in the induced channel current of the MOS device) of the piezoelectric polymer. The MOS part of the POSFET taxel is obtained by using the n-MOS technological module of a non standard CMOS (Complementary Metal Oxide Semiconductor) technology, based on 4 μm p-well ISFET (Ion Sensitive)/CMOS process. The MOS has interdigitated gate structure and the aspect ratio (W/L) of around 625 (W = 7,500 μm and L = 12 μm).

There are many challenging issues before these chips can be integrated as robot's hand skin, including (1) limited space availability on robot's hands and fingertips, and (2) process variability (uncertainty and spread of the transistor technological and geometrical parameters). Such difficulties can be circumvented by using innovative circuit design. As a first step, before designing the readout circuit, there is need to analyze the POSFET bias circuit schema as this greatly affects the overall readout circuit performance. To this aim, a POSFET bias circuit configuration and the graphical-aided methodology to treat the design are proposed in this work. In this context, this paper extends our previous research activity on POSFETs tactile sensing devices [1–3].

2 POSFET Bias Configuration and Proposed Methodology

Rather than conventional common-source configuration of MOSFET and large bias gate resistor [4], we investigated the common-drain configuration, as shown in Fig. 32.2, for biasing the POSFET devices. The choice of this configuration stems from following advantages: (1) it uses minimal number of components that results in reduced size of the readout circuitry, thus making it more suitable for the target application, (2) the configuration is best suited against the statistical variations of the transistor technological-geometrical parameters.

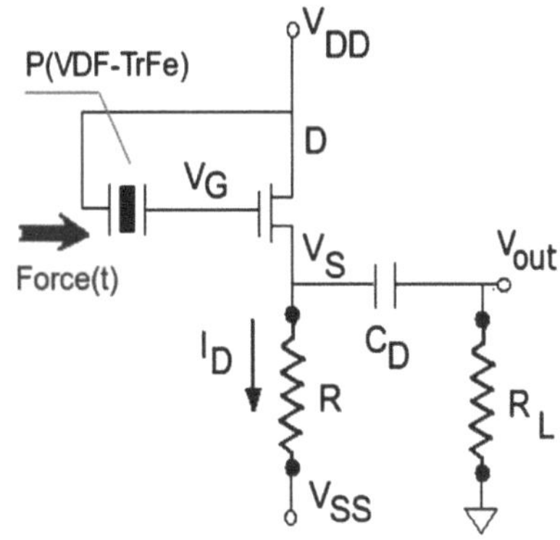

Fig. 32.2 The POSFET common drain configuration with bias resistance R between source and negative supply voltage Vss, C_D provide AC signal coupling and R_L models the input resistance of the following stage

For a common drain configuration transistor, as in Fig. 32.2, the small signal gain in the mid-frequency range for $R << R_L$ is given by:

$$A_v = \frac{V_{out}}{V_G} = \frac{g_m R}{1 + g_m R} \tag{32.1}$$

The unit-gain, which becomes independent of technological-geometrical parameters values, is met for R and g_m values that hold the condition expressed as $g_m R >> 1$(in practice $g_m R \geq 10$ is enough to obtain unity gain i.e. $\approx 1\ V/V$). The problem to overcome is to select the resistance value that assures such condition and the main difficulty lies in the fact that in the floating gate MOS configuration (as the one of the POSFET device) the gate voltage cannot be estimated or controlled beforehand. Thus, this work is aimed to furnish a criterion, by means of a graphical diagram, to select the R value (and consequently the MOS transistor channel current I_D) that ensures $g_m R \geq 10$.

The proposed graphical-aided methodology combines the MOS transistor g_m/I_D characteristics, which is commonly used to treat analog design synthesis [5, 6], and the ACM MOS compact model [7] in order to effectively explore the bias design space defined by the couple of values $\{R, V_s\}$.

The Eq. 32.2 below better describes the condition $g_m R \geq 10$ (but equivalent) for ensuring $A_v \approx 1\ V/V$:

$$g_m R = g_m R \frac{I_D}{I_D} = \frac{g_m}{I_D} R I_D \geq 10 \Rightarrow \frac{g_m}{I_D} \geq \frac{10}{R I_D} = \frac{10}{(V_s - V_{ss})} \tag{32.2}$$

By using the ACM MOS model it is possible to write the g_m/I_D parameter as function of I_D as follows:

$$\frac{g_m}{I_D} = \frac{1}{n U_T} \frac{2}{\sqrt{1 + \frac{I_D}{I_S}} + 1} \tag{32.3}$$

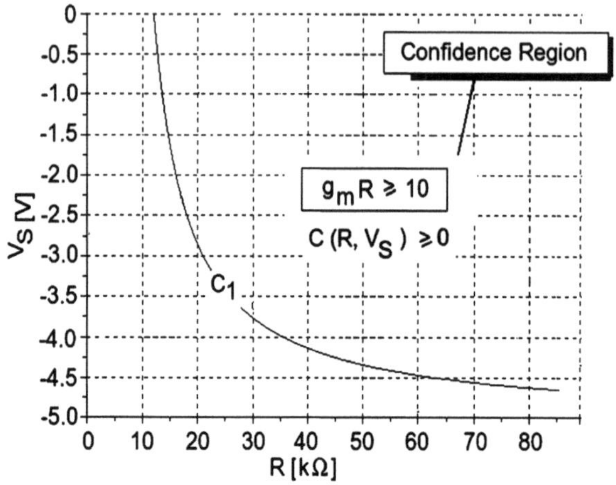

Fig. 32.3 Confidence region

Where $I_s = \frac{1}{2} n \mu_n C'_{ox} \frac{W}{L} U_T^2$ can be experimentally measured. Moreover, as $I_D = \frac{V_s - V_{ss}}{R}$, we achieve the inequality (32.4), with variables$\{R, V_s\}$.

$$C(R, V_s) = \frac{1}{nU_T} \frac{2}{\sqrt{1 + \frac{V_s - V_{SS}}{RI_S}} + 1} - \frac{10}{V_s - V_{SS}} \geq 0 \tag{32.4}$$

Thus, it is possible to define a confidence region (see Fig. 32.3) where all values $\{R, V_s\}$ that fall into the confidence region guarantees $A_v \approx 1\ V/V$. This region is bounded by the contour C_1 defined by the points $\{R, V_s\}$ for which the inequality (32.4) becomes equal to zero.

3 Methodology: Case Study

In this case study, we add more constraints such as C_2 that represents the points $\{R, V_s\}$ for $V_{GS} = V_{DS}$ (i.e. maximum possible value for V_{GS}), as well as the maximum R and I_D values.

We consider $V_{DD} = 0$, $V_{SS} = -5$ V, $R_{max} = 50$ kΩ and $I_{D,max} = 200$ μA. The MOS parameter I_s has been extracted by means of empirical MOS characterization and it results $I_s = 0.55$ μA. The confidence region is then bounded by the curves C_1 and C_2 and the maximum constraint R and I_D (see curve C_3) values as Fig. 32.4 shows. The confidence region condition usage is as follows: we select a value of R to be used as bias resistance, then the voltage V_S is measured. After that, we can select a R value for which the POSFET bias point (or $\{R, V_s\}$ point) are within the confidence region and compliant with the imposed constraints.

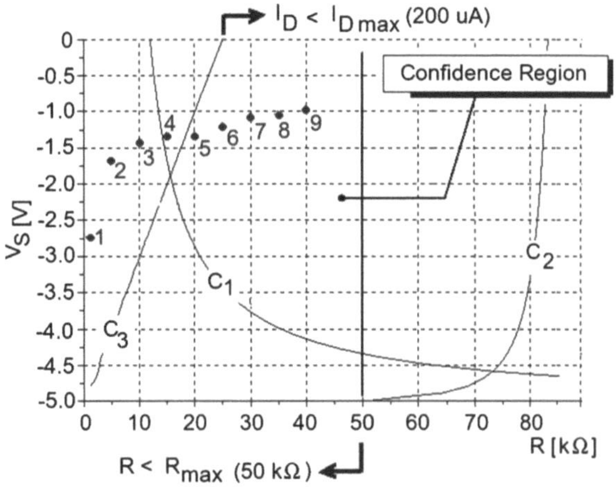

Fig. 32.4 Confidence region for the case study. Points' values are in Table 32.1

Table 32.1 Experimental results. For each *R* value, it has been measured *Vs* and I_D, i.e. the bias point. The rest of the parameters are calculated

Point	R [kΩ]	I_D [μA]	Vs [V]	g_m/I_D [V^{-1}]	$(Vs - VSS)^{-1}$ [V^{-1}]	A_v [V/V]
1	1	2,300	−2.73	1.18	0.441	0.73
2	5	659.3	−1.69	2.15	0.302	0.88
3	10	356.0	−1.44	2.90	0.281	0.91
4	15	244.7	−1.33	3.48	0.272	0.93
5	20	183.1	−1.34	4.00	0.273	0.94
6	25	151.5	−1.21	4.36	0.264	0.94
7	30	130.8	−1.08	4.69	0.255	0.95
8	35	112.8	−1.05	5.01	0.253	0.95
9	40	100.7	−0.97	5.31	0.248	0.96

Table 32.1 summarizes the achieved bias point for different resistance values shown in Fig. 32.4. Then, a bias resistance value higher than 20 kΩ (which corresponds to the point 5) should be used.

4 Conclusions

The methodology presented in this work optimizes the design of the proposed floating gate POSFET bias circuit. The presented approach helps to select the POSFET operating point, which is the first step in the design of read out circuitry. The methodology has been validated with measurements on actual POSFETs. Following this procedure, one can save a considerable number of trials and simulations.

Acknowledgment This work is supported by the European Commission Project ROBOSKIN (Grant No. ICT-FP7-231500). The research leading to these results has received funding from

European Community's 7th framework "People" specific programme – Marie Curie Actions – under grant agreement PCOFUND–GA–2008–226070.

References

1. Dahiya RS, Metta G, Valle M, Adami A, Lorenzelli L (2009) Piezoelectric oxide semiconductor field effect transistor touch sensing devices. Appl Phys Lett 95(3):34105
2. Barboni L, Dahiya RS, Metta G, Valle M (2010) Interface electronics design for POSFET devices based tactile sensing systems. In: Proceedings of the IEEE – ROMAN 2010, Viareggio, pp 1–6
3. Dahiya RS, Valle M, Metta G, Lorenzelli L, Adami A (2009) Piezo-polymer-FET devices based tactile sensors for humanoids robots XIV. Annual conference on sensors and microsystems AISEM 2009, Pavia, 24–26 Feb 2009
4. Weller HJ, Setiadi D, Binnie TD (2000) Low-noise charge sensitive readout for pyroelectric sensor arrays using PVDF thin film. Sensor Actuator 85(1–3):267–274
5. Barboni L, Fiorelli R, Silveira F (2006) A tool for design exploration and power optimization of CMOS RF circuits blocks. IEEE International Symposium on Circuits and Systems (ISCAS), Island of Kos, 21–24 May 2006, ISBN: 0-7803-9389-9
6. Jespers PGA (2007) The Gm/ID design methodology for CMOS analog low power integrated circuits. Springer, New York/London. ISBN 978-0-387-47101-6
7. Galup-Montoro et.al (2007) The advanced compact MOSFET (ACM) model for circuit analysis and design. In: Proceedings of the custom integrated circuits conference IEEE, San Jose, 16–19 Sep 2007, pp 519–526, ISBN: 978-1-4244-1623-3

Chapter 33
Micro-Power Scavenging from Multiple Heterogeneous Piezoelectric and RF Sources

Aldo Romani, Alessandra Costanzo, Diego Masotti,
Enrico Sangiorgi, and Marco Tartagni

Since power harvesting applications are often constrained by the low levels of power available from individual energy transducers, it is essential for energy converters to efficiently deal with multiple independent and heterogeneous sources. This paper will present two actively controlled power conversion schemes able to deal with multiple piezoelectric and radio-frequency (RF) energy sources. Both converters are based on active control and make use of an ultra-low power standard microcontroller unit. The intrinsic power consumption of the harvesters are respectively 5.5 and 6.8 μW per source. The power harvesters were characterized with commercial piezoelectric transducers and with a custom designed rectenna. The achieved values of harvested power of tens of μW show that active control boosts performance of at least +41% as a worst case at the expense of a negligible intrinsic power consumption.

1 Introduction

Energy supply is currently one of the most limiting factors to the deployment of pervasive electronic systems. Since power consumption of electronic devices is decreasing, small amounts of energy scavenged from the environment may satisfy the energy budget of a micro-system. However, since in many realistic environments the available power is as low as few $\mu W/cm^2$ [1], the possibility of harvesting energy from multiple heterogeneous type of sources would make the break-even point with power requirements closer. This paper presents two

A. Romani (✉) • A. Costanzo • E. Sangiorgi • M. Tartagni
Department of Electronics, Computer Science and Systems, University of Bologna, Cesena, Italy
e-mail: aldo.romani@unibo.it

D. Masotti
Department of Electronics, Computer Science and Systems, University of Bologna, Bologna, Italy

A. D'Amico et al. (eds.), *Sensors and Microsystems: AISEM 2011 Proceedings*,
Lecture Notes in Electrical Engineering 109, DOI 10.1007/978-1-4614-0935-9_33,

schemes of power converters for harvesting power respectively from multiple independent piezoelectric transducers and from RF energy belonging to multiple radio-frequency bands.

2 Energy Extraction from Multiple Piezoelectric Transducers

In [2] we showed how the synchronous extraction of charge could greatly increase the output power from piezoelectric transducers subject to random vibrations. In [3] we extended the technique to the case of multiple independent transducers. As detailed in [3] charge is extracted from the transducers by activating different resonant circuits in correspondence of voltage peaks and temporarily storing energy in an inductor. Figure 33.1 shows the structure of the converter.

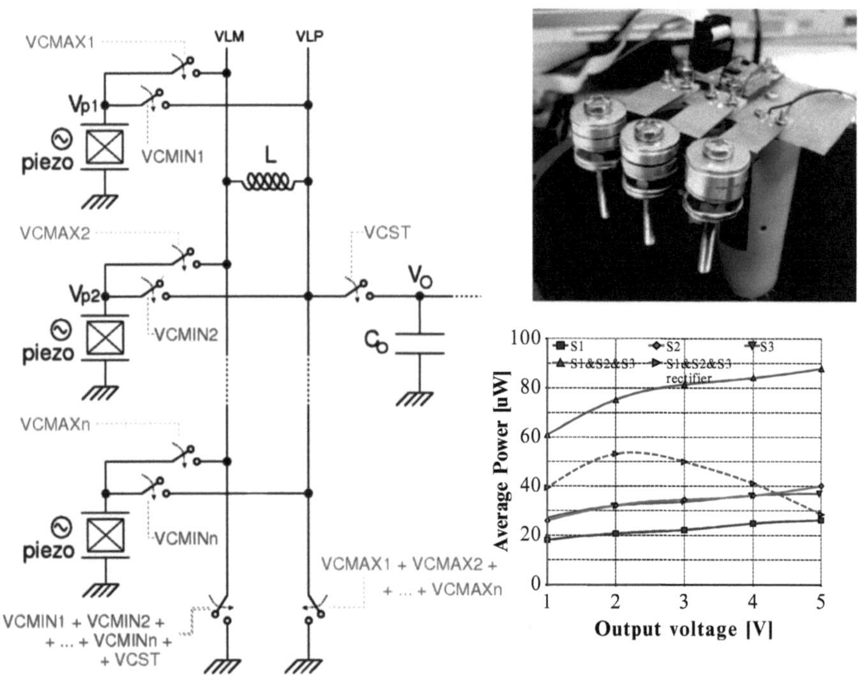

Fig. 33.1 Structure of the synchronized switching energy harvester for multiple piezoelectric transducers. During all measurements three Piezo Systems Q220-A4-303YB transducers were stimulated with realistic vibrations by a custom shaker system. The energy harvested by the proposed converters and by a passive rectifier in presence of a set of 0.11 g vibrations acquired on a train passenger cabin is also reported as a function of the bias point

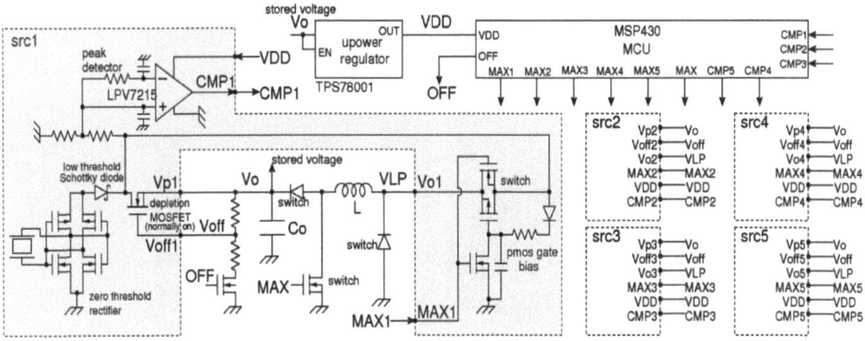

Fig. 33.2 Self-powered circuit implementation of the multi-source piezoelectric converter. Up to five piezoelectric sources with no specific constraints can be handled. A passive interface is used for starting up the system and is cut off when the microcontroller is activated

A self-powered circuit implementation of this energy harvester was designed (Fig. 33.2). The core is a MSP430F122 microcontroller. The analog switches are implemented either with discrete MOSFETs or with BAS70 Schottky diodes. A passive start-up circuit consisting in five zero-threshold rectifiers was used for initially storing charge in the storage capacitor C_O. When V_O reaches about 2.4 V the micro-power voltage regulator turns on and supplies $V_{DD} = 2.2$ V to all the active components. As the MCU turns on, the OFF signal is activated for cutting the passive interface off. Five comparators detect voltage peaks on the transducers. In correspondence of these events, an interrupt request wakes up the MCU from a 0.7 μA stand-by mode for driving switches and transferring energy from the transducer to the storage capacitor. The available output power obtained with three small sized transducers (Piezo Systems Q220-A4-303YB) subject to a set of train vibrations with $a_{RMS} = 0.11$ g is reported in Fig. 33.1.

3 Energy Extraction from Multiple RF Bands

The second proposed converter is composed of a multi-resonator receiving antenna [4] matched to a one-stage rectifier using low threshold Schottky diodes. The antenna is tuned for the GSM900, GSM1800 and Wi-Fi frequency bands. A picture of the rectenna assembly is shown in Fig. 33.3.

As detailed in [5], a good trade-off for the rectenna load generally occurs when V_{RECT} is kept in vicinity of the halved open-circuit DC output voltage. A solution is shown in Fig. 33.4. The power converter periodically samples and halves the open-circuit voltage. A comparator generates Φ and $\bar{\Phi}$ for keeping V_{RECT} around the reference voltage. The RF harvester was tested as shown in Fig. 33.5. During tests the measured RF-DC conversion efficiency was $\eta_{\text{\_RF-DC}} = 31\%$ while the DC-DC

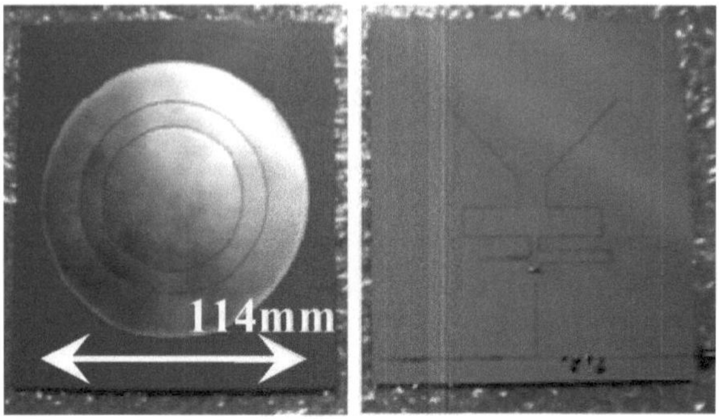

Fig. 33.3 Prototype of the multi-band rectenna device for GSM900, GSM1800 and Wi-Fi bands

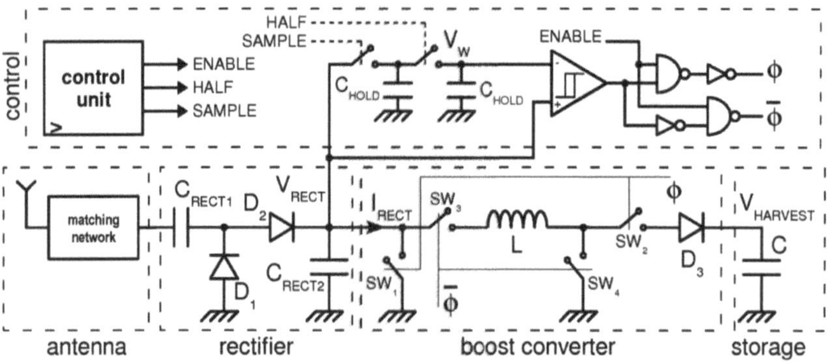

Fig. 33.4 Scheme of the MPPT converter for RF energy harvesting from a rectenna device

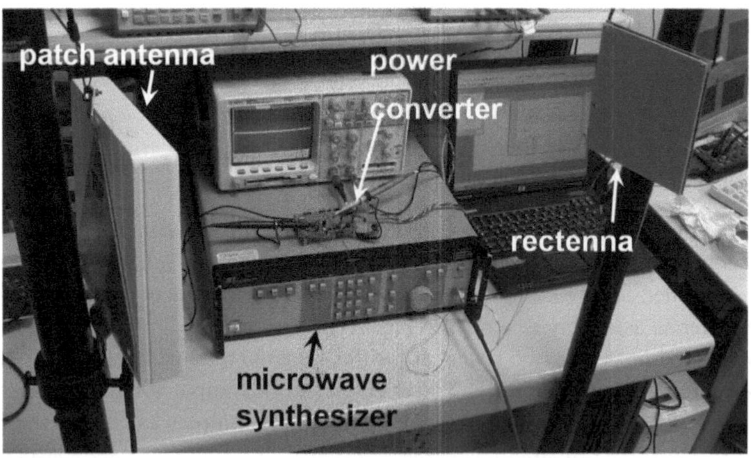

Fig. 33.5 Experimental setup for the multi-band RF harvester: a patch antenna was used for generating RF signals at 900 and 1,800 MHz at given power levels and distance

Table 33.1 Power consumption of the multi-source piezoelectric energy harvester

Component	Power (μW)
Passive interface cut-off circuit	0.4
LDO regulator (quiescent + load current)	9.8
Peak detectors	4.1
Dynamic switching power	0.1
MSP430 microcontroller	3.3
Total	17.7

Table 33.2 Power consumption of the RF power converter

Component	Power (μW)
MSP430 in LPM3 mode with Timer active	2.3
Comparator	1.5
Dynamic switching power	3.0
Static power consumption of switches	<0.1
Total	6.8

converter efficiency η_{CONV} ranged between 78% and 87%. At 900 MHz an output power $P_O = 59\ \mu W$ was obtained with an impinging RF power $P_{AV} = 132.7\ \mu W$, while at 1800 MHz $P_O = 8.15\ \mu W$ with $P_{AV} = 23.6\ \mu W$.

4 Conclusion

The advantages of active converters become evident when the extra harvested power, with respect to passive interfaces, is higher than their intrinsic power consumption. This is true for both the proposed micro-power converters. The reported main contributions to intrinsic power consumption of the two proposed converters were estimated in typical operating conditions. The control of both converters was based on an ultra-low power MSP430 MCU from TI. The results are reported in Tables 33.1 and 33.2. The proposed circuits showed to consume just few μW per source during operation and demonstrate the feasibility and the higher efficiency of actively controlled power harvesting.

Acknowledgments The research leading to these results has received funding from the European Community's Seventh Framework Programme (FP7/2007-2013) under grant agreement NANOFUNCTION no. 257375 and from the ENIAC Joint Undertaking under grant agreement END no. 120214. The authors thank Fondazione Cassa dei Risparmi di Forlì for financial support.

References

1. Vullers R, van Schaijk R, Doms I, Van Hoof C, Mertens R (2009) Micropower energy harvesting. Solid-State Electron 53(7):684–693
2. Romani A, Tamburini C, Paganelli RP, Golfarelli A, Codeluppi R, Dini M, Sangiorgi E, Tartagni M (2009) Actively controlled power conversion techniques for piezoelectric energy harvesting applications. AISEM 2009 proceedings. Lecture notes in electrical engineering, vol 54, pp 359–363
3. Romani A, Paganelli R, Tartagni M (2010) A scalable micro-power converter for multi-source piezoelectric energy harvesting applications. Procedia Engineering. Eurosensors XXIV, vol 5, pp 782–785
4. Rizzoli V, Bichicchi G, Costanzo A, Donzelli F, Masotti D (2009) CAD of multi-resonator rectenna for micro-power generation.EuMC 2009, Rome, pp 1684–1687
5. Costanzo A, Fabiani M, Romani A, Masotti D, Rizzoli V (2010) Co-design of ultra-low power RF/microwave receivers and converters for RFID and energy harvesting applications. Microwave Symposium Digest (MTT), 2010 IEEE, pp 856–859

Chapter 34
Wireless Energy Meters for Distributed Energy Efficiency Applications

Grazia Fattoruso, Ciro Di Palma, Saverio De Vito, Valentina Casola, and Girolamo Di Francia

European and national statistics on energy consumptions show that buildings have a significant energy impact. Their consumptions approximately amounts 40% of the total energy and, according to forecasts, they trend up during next years. It is known that the application of energy smart metering tools can boost energy savings in buildings, leveraging on enhanced awareness. This work presents a pervasive power usage monitoring system based on a wireless energy meter network which can be easily deployed to monitor energy consumptions of appliances in households or computing hardware and related infrastructures in data centers. Two wireless energy meters, as base units of a sensor network, has been designed and developed: a power adapter energy meter and a clamp based energy meter. The adapter is to be employed for monitoring devices that can be plugged to a power outlet while the clamp for heavy loads and devices that cannot be safely or easily unplugged. A base station receives data gathered through all sensors of the network, acting as a gateway to the internet. Ad-hoc web based GUIs provide users with relevant information about real time and aggregated energy consumptions in the selected application.

1 Introduction

In December 2008, the EU adopted an integrated energy and climate change policy including ambitious targets for 2020. It hopes to address Europe towards a sustainable future with a low-carbon, energy-efficient economy by cutting greenhouse

G. Fattoruso (✉) • S. De Vito • G. Di Francia
Basic Materials and Devices Department, ENEA – National Agency for New Technologies, Energy and Sustainable Development, Portici (NA), Italy
e-mail: grazia.fattoruso@enea.it

C. Di Palma • V. Casola
Computer Science and Systems Department, University of Napoli Federico II, Naples, Italy

A. D'Amico et al. (eds.), *Sensors and Microsystems: AISEM 2011 Proceedings*, Lecture Notes in Electrical Engineering 109, DOI 10.1007/978-1-4614-0935-9_34,

gases by 20%; reducing energy consumption by 20% through increased energy efficiency; meeting 20% of our energy needs from renewable sources. This policy action has drawn the attention of all people, from scientific community to common citizens, energy service and structure owners, into energy issues. Regarding energy saving issue in buildings, extensive studies and statistics on energy consumptions have showed that buildings (residential, commercial, etc.) have a significant impact. They are responsible for 40% of the total energy (70% of the electricity) against 32% of industry and 28% of transport and, according to forecasts, their energy demand will trend up during the next years [1].

Just as an example, electricity consumption for hardware and infrastructure in data centers has been steadily growing over the past years in the EU. Currently it amounts 40 TWh/y and it will double next years, unless specific energy efficiency measures are going to be implemented [2, 3]. Moreover, the electricity demand of residential buildings approximately amounts 22% of the total European electricity consumption. Most of this demand is spent for keeping comfortable the indoor climate (12% for cooling and 31% for heating) [1].

Specific studies have picked out that the consumers, from the common citizens to the company owners and public decision-makers, are not provided with effective feedback about the energy consumptions and hence the awareness of existing energy saving potentials is strongly limited. Actually, appropriate tools and technical know-how to enhance energy savings potentials are still in their infancy, while their diffusion is limited to a small number of early birds [4–6]. Consequently, a sufficient monitoring of energy consumption in buildings is currently quite rare. Nevertheless, it is known that a pervasive and continued power monitoring in data centers as well as in households allows to achieve significant energy savings (between 10% and more than 15%) [2, 6, 7].

In this work, we present a pervasive power usage monitoring system based on a wireless energy meters network for appliances in households or computing systems and related infrastructures in data centers. Two wireless energy meters, as base units of the sensor network, have been designed and developed: a power adapter energy meter for devices that can be plugged and a clamp based energy meter for heavy loads and devices that cannot be easily or safely unplugged. A base station receives data gathered through all sensors of the network, acting as a gateway to the internet. Ad-hoc web based GUIs provide users with relevant information about real time and aggregated energy consumptions in the selected energy efficiency application.

2 Proposed Pervasive Architecture

The deployment of intelligent sensing technology is being crucial in the realization of energy-efficient buildings. In a smart metering approach for energy savings, accurate and detailed energy usage information by one or more appliances is

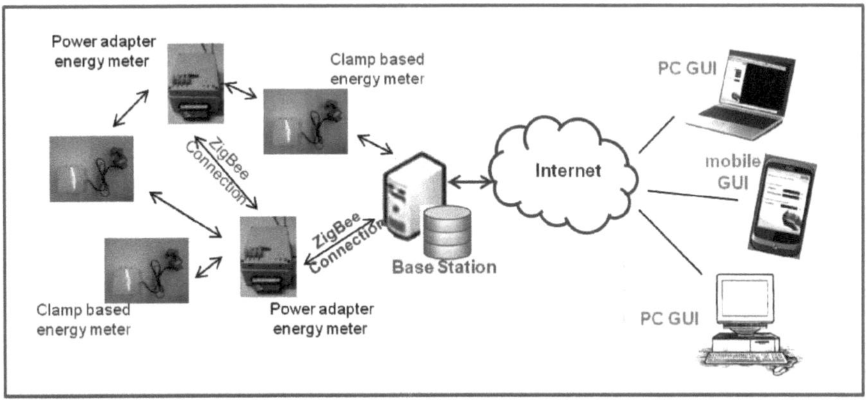

Fig. 34.1 The architecture of the proposed pervasive energy usage monitoring system

gathered through intelligent sensor nodes built up by energy meters connected to a central unit by network communication.

In this work, we present an integrated and pervasive energy usage monitoring system developed for distributed energy efficiency applications in buildings, especially in data centres. This system consists in a wireless intelligent sensor network realised by ZigBee-compliant and mesh (multi-hop) based topology. The base unit of the network is a multi-sensor node built up by an energy meter and two temperature and humidity sensors.

The multi-sensor nodes and the mesh topology provide the proposed system with self-healing and self-configuring capabilities, and scalability to several monitoring scenarios (Smart Building, Smart Office/Home Office, Data Center).

The real time power, temperature and humidity data gathered by sensor nodes are sent via wireless connection to a base station, which acts as gateway to the internet. These data, eventually stored in a database, are mining and displayed through ad-hoc web based GUIs, running on PC, smart phone or tables.

2.1 Base Station and Smart Energy Meters

The proposed pervasive architecture consists in three main components: a base station, two type of smart energy meters and web based GUIs (Fig. 34.1).

In order to monitor the power usage of any type of electric devices from the racks in the data centers to the washers in the households, two type of energy meters have been designed and developed: a power adapter and a clamp based smart energy meters. The first one can be used for measuring energy consumptions of devices that can be plugged to a power outlet; the second one has been designed specifically for heavy loads and devices that cannot be easily or safely unplugged.

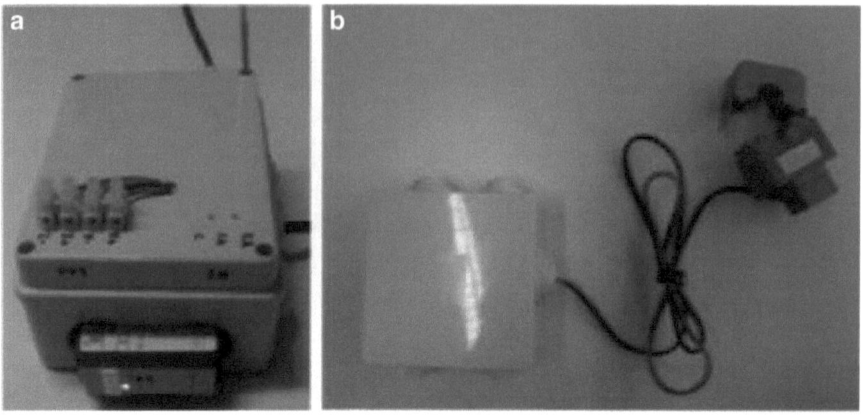

Fig. 34.2 The developed energy meters. (**a**) Power adapter energy meter. (**b**) Clamp based energy meter

Table 34.1 Electrical features of the developed energy meters

Parameter	Power outlet adapter energy meter	Clamp based energy meter
Input signal	0–5A	0–50A
Output signal	0–4 V DC	0–5 V DC
Load resistance	12 mΩ	≥1 MΩ
Supply voltage	220 V	Self-powered
Response time	300 ms	<300 ms
Frequency bandwidth	600 Hz	50/60 Hz

The power adapter energy meter is packaged in an easy-to-install format (Fig. 34.2a); the appliance to be monitored can be simply plugged to this energy meter device, which in turn can be connected into AC power outlet.

The device is basically built up around an universal converter for true RMS (Root Mean Square) current measurement and a TelosB Mote TPR2420 upon which the temperature and humidity sensors are integrated. Embedded software components, coded in NesC, allow local data acquisition and processing while the runtime support of the open-source operating system, TinyOS, provides base functionalities as network formation, packets routing and distributed management of the mesh topology.

The clamp based energy meter is built around a current clamp capable to measure the current on electrical devices (Fig. 34.2b). It is built up by split-core AC current transducer LEM AT-B5 connected to a TelosB Mote TPR2420.

The main electrical features of the two energy meters are listed in the Table 34.1.

The base station of the proposed architecture acts as data sink, storing the data sent by energy meters; performs the function of re-broadcasting transmitting on the TCP/IP port the packets received by a specific node of the network connected via USB; stores the received data on open source database; manages the data access by the web based interfaces.

Ad-hoc GUIs have been designed and implemented for making easily visible to the users the real time and aggregated energy consumptions, also in term of the updated distance from selected saving goals. This type of information would enable the users to know what to change in their behaviors and practices for better to achieve energy efficiency goals. Mobile GUIs for tablets and smart phone will be developed in order to facilitate the buildings energy usage monitoring.

3 Conclusions

Two type of wireless energy meters as base units of a pervasive energy consumption monitoring system have been designed and developed in order to provide both citizens and technicians with appropriate tools to optimize respectively the energy usage in households and data centers. The proposed system will be tailored to real households and data centers and mobile and user-friendly GUIs will be integrated in the proposed architecture for an always at hand solution.

References

1. Crema L (2010) Sustainable energy system & IT. Fondazione Bruno Kessler Workshop: Il Futuro del silicio nel fotovoltaico, October. http://www.fbk.eu/it/futuro-silicio-fotovoltaico
2. IEE Programme (2009) E-Server project, "Efficient Server technology. Energy and Cost savings in the data centre", February. www.efficient-server.eu
3. Pacific Gas and Electric Company (2006) High performance data centers – a design guidelines sourcebook
4. Björkskog AC, Jacucci G, Gamberini L, NieminenT, MikkolaT, Torstensson C, Bertoncini M (2010) EnergyLife: pervasive energy awareness for households. In: Proceedings of the 12th ACM international conference adjunct papers on Ubiquitous computing (Ubicomp'10). ACM, New York, pp. 361–362
5. Duarte LFC, Zambianco JD, Airoldi D, Ferreira EC, Siqueira Dias JA (2011) "Characterization and break down of the electricity bill using custom smart meters: a tool for energy-efficiency programs". Int J Circuit Syst Signal Process 5(2):116–122
6. Decorme R (2010) CIP-ICT-PSP-2009 Project: energy efficiency in European social housing. Technical solutions, January. http://www.e3soho.eu/spip.php?article26
7. Firth S, Lomas K, Wright A, Wall R (2008) Identifying trends in the use of domestic appliances from household electricity consumption measurements. Energy Buildings 40:926–936

Chapter 35
Mass Response of A CMOS-Compatible, Magnetically Actuated MEMS Microbalance

V. Russino, F. Pieri, and A. Nannini

In this work, the mass response of a resonant, CMOS (Complementary MOS) compatible MEMS sensor, oriented at the detection of diagnostic markers, is presented. The sensor is fabricated with a MEMS (Micro-Electro-Mechanical System) post-processing method on a standard, CMOS-based VLSI technology, retaining maximum compatibility with the CMOS process flow. The mechanical resonator is based on inductive actuation and detection, and the sensing is based on the microbalance principle. A protocol for covalent bonding of organo-functional silanes (to be used as link sites for biomolecular probes) on the resonator surface is presented. The effect on the mechanical frequency response of a test mass attached to the surface is demonstrated by grafting of gold nanoparticles (NPs) to the amino-terminated surface silanes. The measured mass sensitivity compares favorably both with standard Quartz Crystal Microbalances (QCM) and with existing MEMS-based approaches.

1 Introduction

MEMS resonators have been proposed as biosensors based on the gravimetric principle. While the fabrication of piezoelectric MEMS resonant sensors (the microscopic counterpart of the QCM) is certainly possible [1], other transduction mechanisms are accessible in the MEMS world. Capacitive MEMS resonant mass sensors with extremely low mass sensitivities have been demonstrated [2], but the transfer of this sensing approach to the biosensing world is not straightforward, given the problems of compatibility between a standard MEMS process/structure

V. Russino (✉) • F. Pieri • A. Nannini
Dipartimento di Ingegneria dell'Informazione, Università di Pisa, Pisa, Italy
e-mail: vincenzo.russino@iet.unipi.it

A. D'Amico et al. (eds.), *Sensors and Microsystems: AISEM 2011 Proceedings*, Lecture Notes in Electrical Engineering 109, DOI 10.1007/978-1-4614-0935-9_35,

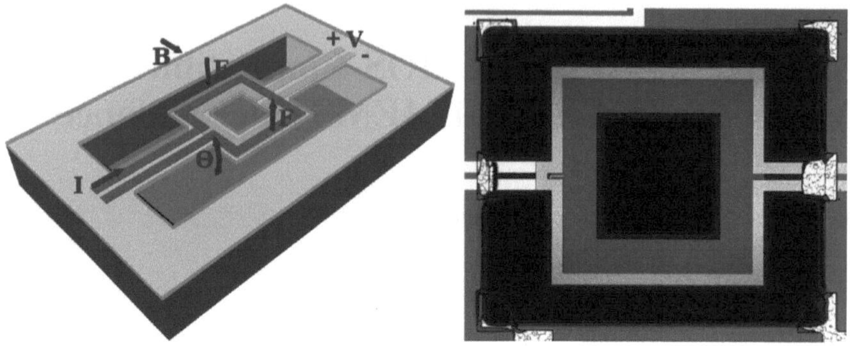

Fig. 35.1 Structure of the fabricated microbalance, with the two embedded inductors in *red* and *light blue* (*left*); optical view of an actual resonator (*right*)

and the steps required for bio-functionalization (i.e. deposition and patterning of the bioactive layer) and operation (i.e. exposure to biological samples).

In this work, we address several of these issues (functionalization, compatibility with on-chip ICs, operation in a liquid environment) by developing a CMOS-based torsional resonator with magnetic driving and sensing. The resonator is based on a dielectric plate with two embedded inductors, suspended to two torsional springs (Fig. 35.1). An external static magnetic field is used for actuation and sensing. The first (driving) inductor is fed a sinusoidal current, which sends the plate into mechanical resonance. The plate oscillation is detected as an induced electromotive force at the terminals of the second (sensing) inductor.

Frequency selectivity is ensured because of low damping losses in the mechanical structure. From the electrical point of view, the device behaves as a two-port resonator.

2 Fabrication

The fabrication of the MEMS component is based on a CMOS-compatible bulk micromachining technology, and is documented elsewhere [3]. Here only the bio-functionalisation protocol, taking place after fabrication of the MEMS structure, is presented. A preliminary cleaning of the sample in an ammonia based solution (NH_4OH(25%):H_2O:H_2O_2(30%) 1:4:1 vol.) is performed. After 5′ in the solution, the sample is rinsed twice in deionized water (D.W.). After this step, the resonator surface is terminated by hydroxyl groups, which are required to allow covalent bonding of the silane. While sulfuric acid-based solutions (e.g. piranha) are most commonly used in the literature [4], they are not compatible with exposed aluminum which is present in our samples at the external electric pads.

Immediately after rinsing, the sample is immersed in an aqueous solution of 3-amino-propyl-triethoxysilane (APTES) (0.05% vol.) for 5′, and rinsed again in D.W.

Fig. 35.2 Sample mounted on its ceramic package. The chip contains several different resonators

The sample then undergoes a curing phase by immersion in acetone for 60′. During this phase, loosely bound silane molecules or silane clusters are removed and the overall quality of the silane layer improves [5].

At this point the resonators, mounted in a standard ceramic package (Fig. 35.2), are electrically bonded to the package pads. A first frequency response of the resonator is measured. To eliminate the influence of ambient humidity (a common interferer for gravimetric sensors), the sample is heated in oven at 120°C for 30′ and immediately moved to a sealed aluminum box for the electrical characterization. The resonators are then loaded with an additional mass to verify their behavior as gravimetric sensor.

The added mass is represented by gold NP's which are known to bind to the exposed amino groups of the silane layer [6]. A 25 μL drop of a 30 nm gold NP colloid (British Biocell International) is deposed on the chip. After 60′, the sample is rinsed in D.W., dried in oven (30′ at 120°C), and the frequency response of the resonators is then measured again. To ensure that the NP anchoring is specific to the silane layer, a reference sample underwent the same procedure with the exclusion of the silanization steps.

3 Results and Conclusions

The frequency response was then measured again to evaluate the effect of the added mass on the resonance frequency. The amount of added mass was estimated by Scanning Electron Microscope (SEM) observation of the resonator at several

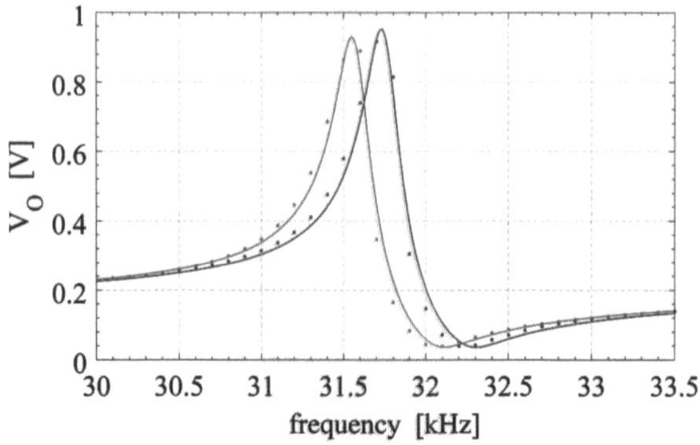

Fig. 35.3 Frequency response of the resonator in Fig. 35.1 before (*blue curve*) and after (*red curve*) exposure to gold NPs. *Dots* represent actual data, while the *curves* are numerically fitted against the data

different test sites, counting the observed NPs and averaging their number for the total resonator surface. A typical frequency response before and after NP's exposure is shown if Fig. 35.3. The resonance frequency after NP exposure is consistently lower than the initial frequency, and corresponds to a frequency shift of a few hundreds Hz. While the resonance frequency tends to drift to higher values after a few days on ambient storage (most likely because of adsorption of ambient moisture), a drying step (30′ at 120°C) recovers its original value almost completely.

The frequency shifts must be compared to the measured values of the added mass to obtain an estimate of the sensor sensitivity. A sensitivity parameter allowing a fair comparison can be defined as in [7]:

$$S = \frac{\Delta f}{f_0} \frac{A}{m}$$

where Δf is the measured frequency shift, f_0 the initial resonance frequency, m the added mass, and A the microbalance surface area. The measured values for our resonators are around 80 m^2/kg, comparing favorably to the typical values for macroscopic Quartz Crystal Microbalances (at about 1 m^2/kg), but also to other MEMS resonant sensors [1, 8].

Acknowledgments The authors thank STMicroelectronics for allowing access to the BCD6s technology. This work was partly financed by the Italian Ministry of Education, University and Research under a PRIN grant.

References

1. Zuniga C, Rinaldi M, Khamis SM, Johnson AT, Piazza G (2009) Nanoenabled microelectromechanical sensor for volatile organic chemical detection. Appl Phys Lett 94:223122
2. Jensen K, Kwanpyo Kim, Zettl A (2008) An atomic-resolution nanomechanical mass sensor. Nature Nanotechnol 3:533–537
3. Paci D, Pieri F, Toscano P, Nannini A (2008) A CMOS-compatible, magnetically actuated resonator for mass sensing applications. Sensor Actuator B 129:10–17
4. Lenci S, Tedeschi L, Pieri F, Domenici C (2011) UV lithography-based protein patterning on silicon: Towards the integration of bioactive surfaces and CMOS electronics Appl Surf Sci 257:8413–8419
5. Kim J, Seidler P, Fill C, Wan SW (2008) Investigations of the effect of curing conditions on the structure and stability of amino-functionalized organic films on silicon substrates by Fourier transform infrared spectroscopy, ellipsometry, and fluorescence microscopy. Surf Sci 602:3323–3330
6. Diegoli S, Mendes P, Baguley E, Leigh S, Iqbal P, Diaz YG, Begum S, Critchley K, Hammond G, Evans S, Attwood D, Jones I, Preece J (2006) pH-dependent gold nanoparticle self-organization on functionalized Si/SiO_2 surfaces. J Exp Nanosci 1:333–353
7. Janshoff A, Galla H, Steinem C (2000) Piezoelectric mass-sensing devices as biosensors – an alternative to optical biosensors? Angew Chem Int 39:4004–4032
8. Shen W, Mathison L, Petrenko V, Chin B (2010) A pulse system for spectrum analysis of magnetoelastic biosensors. Appl Phys Lett 96:163502

Chapter 36
Acoustic Particle Velocity Sensors Based on a Thermal Principle

M. Piotto, P. Bruschi, and F. Butti

A new integrated sensor for the acoustic particle velocity measurement is proposed. The device is based on a thermal principle and is made up of two polysilicon heaters placed over suspended dielectric membranes. The sensor is fabricated by means of a post-processing technique applied to chips designed with a commercial CMOS process. Preliminary measurements confirm the device suitability for acoustic applications.

1 Introduction

Micromachining techniques allowed the miniaturization of thermal flow sensors fabricating devices characterized by low thermal capacities with response times of a few milliseconds. This has opened up possibilities for applications in the acoustic field. In 1996, de Bree et al. [1] proposed the μ-flown, a device consisting in three microwires placed over a suspended dielectric bridge capable of measuring the particle velocity components of an acoustic wave. This device has a few interesting characteristics that distinguish it from pressure microphones. The most important are the absence of a low frequency cut-off and the sensitivity to the direction of the sound wave. In these last years the μ-flown has been developed and a complete three-dimensional sound intensity sensor integrated on a single chip has been recently proposed [2]. However, the μ-flowns presented so far are obtained with a dedicated technological process, not compatible with standard integrated circuit technology. This prevents the integration of the sensor and the read-out electronic

M. Piotto (✉)
CNR IEIIT, Pisa, Italy
e-mail: massimo.piotto@cnr.it

P. Bruschi • F. Butti
Dipartimento di Ingegneria dell'informazione, Università di Pisa, Pisa, Italy

A. D'Amico et al. (eds.), *Sensors and Microsystems: AISEM 2011 Proceedings*,
Lecture Notes in Electrical Engineering 109, DOI 10.1007/978-1-4614-0935-9_36,

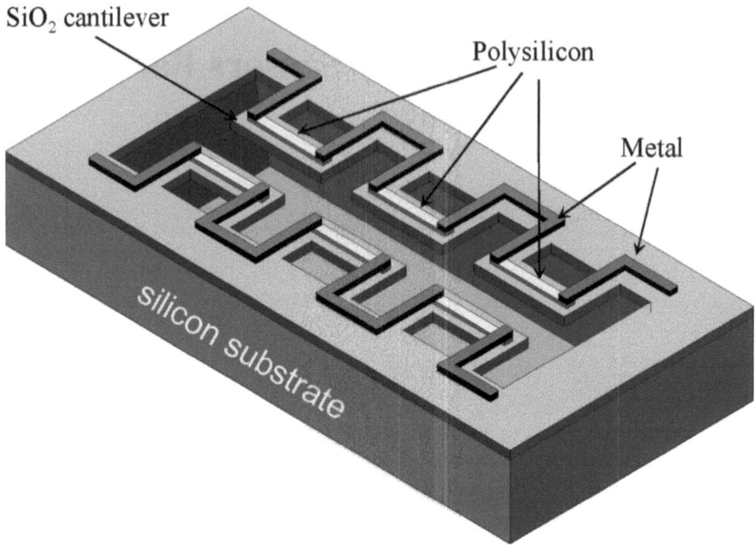

Fig. 36.1 Perspective view of the proposed device

circuits on a single very compact chip. Recently, we have proven that standard thermal flow sensors fabricated by means of a CMOS compatible post-processing technique could be used for dynamic particle velocity detection even if the upper band limit was too low for audio applications [3, 4].

In this work we proposed a new device consisting in two polysilicon wires placed over suspended dielectric membranes. In order to improve the mechanical robustness, each one of the two wires has been divided into different sections, suspended on mechanically independent "U" shaped membranes, as schematically shown in Fig. 36.1.

2 Device Fabrication

The device has been designed with the BCD6s process of STMicroelectronics and a post-processing technique based on a TMAH silicon anisotropic etching has been applied in order to thermally insulate the wires from the substrate. A photograph of the proposed device before the post-processing is shown in Fig. 36.2.

In order to access the bare silicon from the chip front side, the dielectric layers were selectively removed by means of a 1 μm resolution photolithographic step and a buffered HF (BHF) solution. Then the silicon was anisotropically etched using a solution of 100 g of 5 wt% TMAH with 2.5 g of silicic acid and 0.7 g of ammonium persulfate. A single etch step of 60 min at 85°C was sufficient to thermally insulate the structures from the silicon substrate. After the silicon etching, the chip was

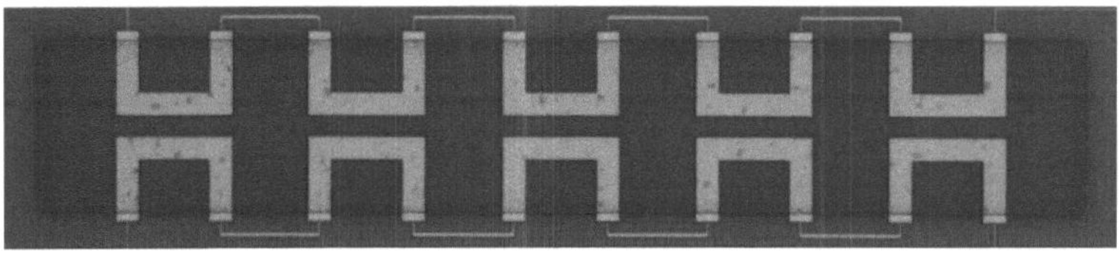

Fig. 36.2 Photograph of the device before the post-processing

glued to a ceramic DIP case by means of epoxy resin and wedge bonding was used to connect selected chip pads to the case pins.

3 Device Characterization

The frequency response of the sensors has been measured using the standing wave tube technique [5]. An excitation loudspeaker was placed at one end of a plastic tube with a diameter of 5 cm and a length of 1 m. A reference microphone was placed at the other closed end. The device was placed inside the tube at 70 cm from the microphone. In order to verify the directivity of the fabricated sensor, a device holder equipped with a precision goniometer has been used. The standing wave tube measurement range is limited by the tube diameter to a maximum frequency of 4 kHz.

If a harmonic excitation at frequency f is applied at the open end of the rigidly terminated tube, the ratio $R(x,f)$ between the particle velocity $u(x,f)$, at any position x inside the tube, and the reference pressure P_{ref}, at the closed end of the tube, is given by:

$$R(x,f) = \frac{u(x,f)}{P_{ref}(f)} = \frac{i}{\rho c} \sin[k(L - x)] \tag{36.1}$$

where k is the wave number, i the imaginary unit, L the tube length and ρ the fluid density. The quantity ρc is usually named characteristic impedance (z_s) and is a fluid constant. The ratio $H(x,f)$ between the output voltage of the device under test (V_D) and the voltage of the reference microphone (V_m) is given by:

$$H(x,f) = \frac{V_D(x,f)}{V_m(f)} = \frac{S_D(f)}{S_m(f)} \frac{u(x,f)}{P_{ref}(f)} = \frac{S_D(f)}{S_m(f)} R(x,f) \tag{36.2}$$

where $S_D(f)$ and $S_m(f)$ are the sensitivities of the device and the microphone, respectively. Assuming that the frequency behaviour of $S_m(f)$ is flat in the investigated frequency range, the amplitude of $H(x,f)$ is proportional to the device sensitivity and the ratio $R(x,f)$. In Fig. 36.4 a plot of the $H(x,f)$ amplitude measured at a fixed x value (i.e. device position) as a function of frequency is shown.

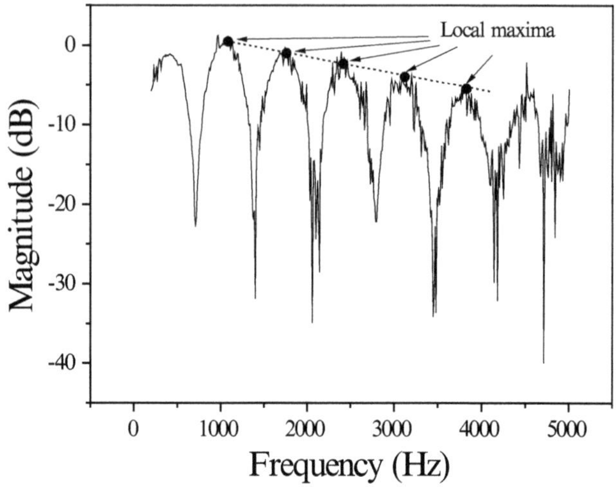

Fig. 36.3 Plot of *H(x,f)* amplitude as a function of frequency; the local maxima are highlighted

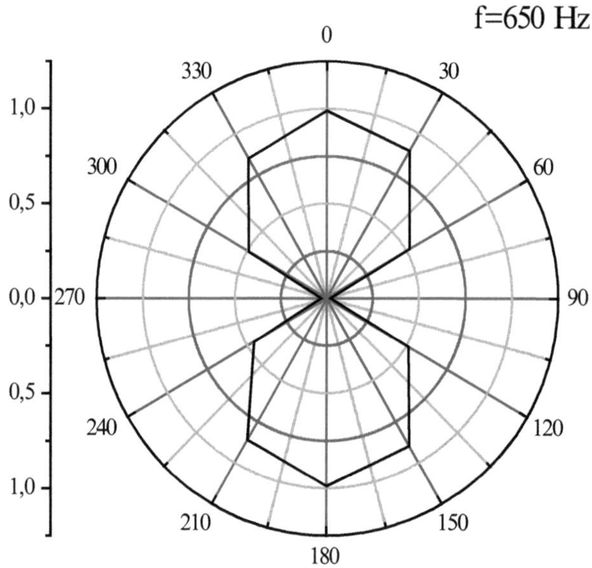

Fig. 36.4 Directional response measured at a frequency of 650 Hz

It can be noted that the sinusoidal dependence on frequency predicted by eq. (1) through the wave number has been obtained up to the maximum frequency of 4 kHz. When the *H(x,f)* amplitude is maximum, the term *R(x,f)* can be considered a constant so the local maxima are proportional to the sensor sensitivity at the respective frequency. In this way, it is possible to obtain the frequency behavior of the sensor sensitivity by simply connecting the maxima as shown in Fig. 36.3.

In our case a slowly decreasing response has been obtained up to the maximum measurable frequency.

As far as the sensor directivity is concerned, Fig. 36.4 shows the polar plot of the normalized response obtained at 650 Hz as a function of the rotation angle. It can be noted that a clear figure-of-eight behavior has been obtained, confirming the dependence of the sensor response on the sound wave direction.

4 Conclusions

A sensor for the acoustic particle velocity measurement has been designed with a commercial CMOS process and has been fabricated by means of a post-processing technique based on a modified TMAH solution. The preliminary measurements have been performed with the standing wave tube technique up to maximum frequency of 4 kHz. In the investigated frequency range, the device has shown a slowly decreasing frequency response and a good dependence on the sound wave direction. Further investigations will be necessary to fully characterized the device behavior.

Acknowledgments The authors thank STMicroelectronics R&D group of Cornaredo (MI) for fabricating the chip.

References

1. de Bree HE, Leussink P, Korthorst T, Jansen H, Lammerink TSJ, Elwenspoek M (1996) The μ-flown: a novel device for measuring acoustic flows. Sensor Actuator A 54:552–557
2. Yntema DR, van Honschoten JW, Wiegerink RJ, Elwenspoek M (2008) A complete three-dimensional sound intensity sensor integrated on a single chip. J Micromech Microeng 18:115004-1–9
3. Bruschi P, Schipani M, Bacci N, Piotto M (2008) A finite element 2-dimensional model for the prediction of the frequency response of thermal gas velocity detectors. Proceedings of the 12th Italian conference on sensors and microsystems, World Scientific, pp. 520–525
4. Bruschi P, Dei M, Piotto M (2009) Frequency response of thermal gas velocity detectors. Proceedings of the 13th Italian conference on sensors and microsystems, World Scientific, pp 331–335
5. de Bree HE (2003) An overview of microflown technologies. Acta Acustica 89:163–172

Part IV
Optical Sensors and Related Techniques

Chapter 37
Static Light Scattering for Measuring Biological Cell Concentration

L. Ciaccheri, A.G. Mignani, A.A. Mencaglia, and L. Giannelli

An experimental study was carried out, aimed at optimizing the optical/geometrical configuration for measuring the concentration of biological cells by means of static light scattering measurements. A LED-based optoelectronic setup making use of optical fibers was experimented, as the precursor of a low-cost device to be integrated in instrumentation for cytometry. Two biological sample types were considered as test samples of the most frequent analyses – cervical cells and urine, respectively. The most suitable wavelengths and detecting angles were identified, and calibration curves were calculated.

1 Introduction

Liquid-based cytology (LBC) is an innovative way of preparing biological samples for cytological examinations in the laboratory. It consists of fixing the collected sample in a preservative alcohol-based fluid for further clarification, centrifugation, and then depositing a thin layer of cells on a slide. The ensuing examination is carried out by the cytologist in the usual way under a microscope. Today, LBC is the most widely used form of its kind in applications to gynaecological cervical smears (PAP-test), for which it was originally developed [1–3]. Subsequently, LBC has progressively gained favour also in many others cytologies [4], especially urinary [5], oral [6], naso-pharyngeal [7], as well as for breast tumor analysis [8].

Since 1999, Hospitex Diagnostics srl has been implementing an innovative and effective proprietary LBC method, the CYTOfast® system [9]. It makes use of

L. Ciaccheri (✉) • A.G. Mignani • A.A. Mencaglia
CNR IFAC, Sesto Fiorentino (FI), Italy
e-mail: l.ciaccheri@ifac.cnr.it

L. Giannelli
Hospitex Diagnostics srl, Sesto Fiorentino (FI), Italy

A. D'Amico et al. (eds.), *Sensors and Microsystems: AISEM 2011 Proceedings*,
Lecture Notes in Electrical Engineering 109, DOI 10.1007/978-1-4614-0935-9_37,

CYTOfast® solution, a universal preservative solution that makes it possible to preserve the physiological structure and morphology of any kind of cell for 24 months at room temperature. Cellular material left in the vial, after the slide preparation, can be used directly for further investigations employing molecular biology techniques (e.g. PCR, hybridization, etc.).

The CYTOfast® system makes use of a *standardization phase* during which a nephelometric reading determines the cellular density of the samples. In accordance with this idea, the system fixes the quantity to add to the slide for every sample, in order to obtain numerically standardized slides, which always contain the same number of cells, distributed as a monolayer, on a spot having a diameter of 17 mm, for a safer, faster, easier and representative screening (approximately 100,000 cells). The better the knowledge of the cellular density, the better the monolayer uniformity and quality and, consequently, the results of cytological analyses will be.

Static light scattering has long been a standard method for cell concentration assessment in biological samples [10]. This paper presents the results of an experimental nephelometric study which was performed on cervical and urine cells in a CYTOfast® solution. The scope of the experiment was to optimize the optical/geometrical configuration for measuring the concentration of biological cells by means of static light scattering measurements. A LED-based optoelectronic setup making use of optical fibers was experimented, as the precursor of a low-cost device to be integrated in instrumentation for cytometric purposes. The most suitable wavelengths and detecting angles were identified, and calibration curves were calculated.

2 Experimental Setup and Measurement Results

Figure 37.1 shows the diagram of the experimental setup. A glass vial containing the biological sample was inserted in a jig. Optical fibers coupled to SELFOC® collimators [11] were used for illumination and detection. Four LEDs were considered for sequential illumination at four wavelengths, i.e. 405, 525, 644, and 850 nm.

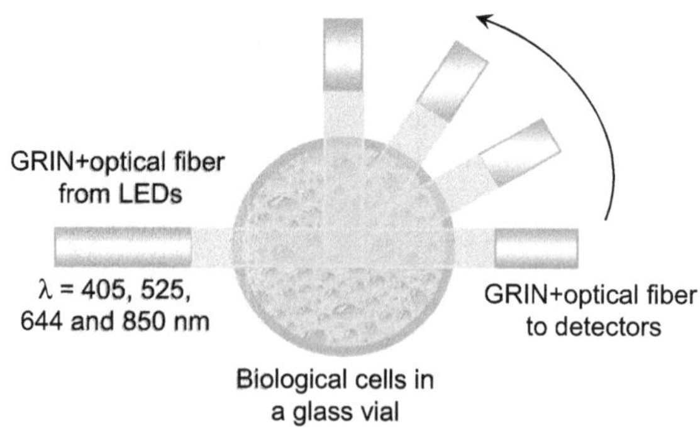

Fig. 37.1 Diagram of the experimental setup

Table 37.1 Summary of sigmoid parameters fitting the experimental data for PAP-test (*left*) and urine (*right*) cell solutions

PAP-test				Urine			
θ (°)–λ (nm)	*A*	*B*	*k*	*θ (°)–λ (nm)*	*A*	*B*	*k*
30–405	1,053	9.9 e^{-2}	1.05	30–405	1,828	1.6 e^{-1}	0.86
30–525	1,032	2.0 e^{-2}	1.04	30–525	1,894	4.9 e^{-2}	0.83
30–644	1,027	3.2 e^{-3}	1.05	**30–644**	**1,842**	**1.1 e^{-2}**	**0.84**
30–850	**1,024**	**1.9 e^{-3}**	**1.11**	30–850	1,770	7.0 e^{-3}	0.89
60–405	1,033	2.3 e^{-2}	0.85	60–405	1,832	1.9 e^{-2}	0.85
60–525	1,023	5.0 e^{-3}	0.85	60–525	2,254	1.9 e^{-2}	0.70
60–644	1,023	7.1 e^{-4}	0.89	60–644	2,217	3.8 e^{-3}	0.73
60–850	1,016	2.8 e^{-4}	0.96	60–850	1,597	4.5 e^{-4}	0.88
90–405	1,018	7.2 e^{-3}	0.84	90–405	1,034	4.6 e^{-9}	2.76
90–525	1,018	1.5 e^{-3}	0.86	90–525	1,786	1.4 e^{-3}	0.84
90–644	1,015	1.3 e^{-4}	0.94	90–644	2,027	3.6 e^{-4}	0.85
90–850	1,017	9.1 e^{-5}	0.98	90–850	1,484	3.0 e^{-5}	1.05

Detection was carried out at 0°, 30°, 60° and 90° with respect to the illumination, so as to measure the light intensity transmitted through the sample, as well as the scattered light at several angles.

Cells from PAP-tests and urine were fixed in CYTOfast® solution, and biological samples were prepared that had cell concentrations in the 20–1,000 cell/mm^3 range, which typically occur in real conditions. While PAP-test solutions were practically colourless, urine solutions exhibited a yellow colour, ranging from pale to intense, depending on the urine concentration.

The entire biological set was optically characterized by measuring the ratio between scattered and transmitted light power, $x = P(\theta)/P(0°)$, at all wavelengths and at all angles. This ratio provided a normalized output that was independent of absorption effects and source intensity fluctuations. The fitting of the experimental curves from which to determine cell concentration as a function of the normalized output, showed a nonlinear behaviour, that was satisfactory represented by means of a sigmoid function, as described by Eq. 37.1.

$$C(x) = \frac{A\,x^k}{B + x^k} \tag{37.1}$$

Table 37.1 summarizes the fitting parameters for all experimented wavelength-angle combinations, and the best conditions are highlighted. The most efficient angle was found to be 30° for both PAP-test and urine cell solutions, while the best wavelength was 850 nm for PAP-test, and 644 nm for the urine. The experimental results and relative fitting functions of PAP-test at 850 nm and urine at 644 nm are shown in Fig. 37.2. Because of their morphological differences, the urine cells exhibited a lower scattering efficiency as compared with the PAP-test cells.

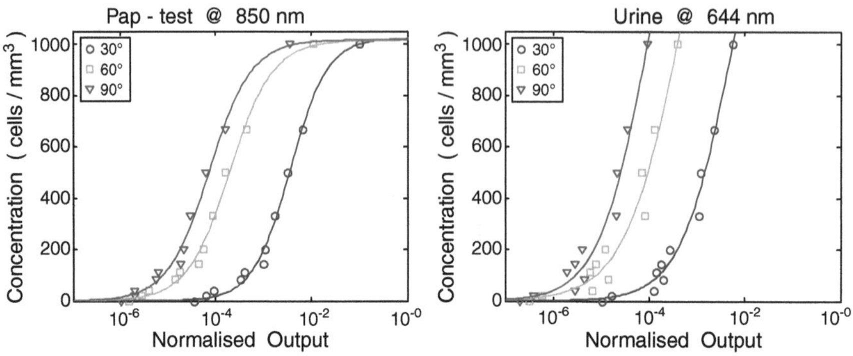

Fig. 37.2 Experimental results (marks) and fitting functions at selected wavelengths for PAP-test (*left*) and urine (*right*) cell solutions

3 Perspectives

PAP-test and urine cells fixed in CYTOfast® solution were considered. A comprehensive nephelometric study for cell density measurements was carried out at several illumination wavelengths and detection angles. This made it possible to determine the best wavelength-angle combination for each type of biological cells. Our study was aimed at implementing an automatic device capable of drawing a fixed cell amount to be smeared on a microscope-glass. In fact, dealing with a fixed number of cells enabled us to obtain a uniform and good quality cell monolayer, which is what is needed for optimal LBC cytological analyses.

Acknowledgments This work was partially funded by the Regional Board of Tuscany, under the DOCUP 1.8.1 funding scheme.

References

1. Saraiya M, Bernard V, Miller J (2010) Liquid-based cytology versus conventional cytology in detecting cervic cancer. J Am Med Assoc 303(11):1034
2. Beerman H, van Dorst EBL, Kuenen-Boumeester V, Hogendoorn PCW (2009) Superior performance of liquid-based versus conventional cytology in a population-based cervical cancer screening program. Gynecol Oncol 112:572–576
3. Siebers AG, Klinkhamer PJJM, Grefte JMM, Massuger LFAG, Vedder JEM, Beijers-Broos A, Bulten J, Arbyn M (2009) Comparison of liquid-based cytology with conventional cytology for detection of cervical cancer precursors: a randomized controlled trial. J Am Med Assoc 302(16):1757–1764
4. Rossi ED, Fadda G (2008) Thin-layer liquid-based preparation of non-gynaecological exfoliative and fine-needle aspiration biopsy cytology. Diagn Histopathol 14(11):563–570
5. Piaton E, Faynel J, Hutin K, Ranchin MC, Cottier M (2005) Conventional liquid-based techniques versus Cytyc Thinprep processing of urinary sample: a qualitative approach. BMC Clin Pathol 5(9):1–7

6. Sato K, Tanaka Y, Yamauchi T, Katakura A, Noguchi S, Fujishiro Y, Takaki S, Tonogi M, Yamane GY (2009) The evaluation of liquid based cytology in oral squamous cell carcinoma. Int J Oral Maxillofac Surg 38(5):565–566, Proc. 19th International Conference on Oral and Maxillofacial Surgery
7. Jan YL, Chen SJ, Wang J, Jiang RS (2009) Liquid-based cytology in diagnosing nasopharyngeal carcinoma. Am J Rhinol Allergy 23(4):422–425
8. Mitsuru K, Yukiko T, Wakaho D, Noriko N, Kosuke Y, Shigeru I (2006) An applicability of liquid-based cytology to fine needle aspiration cytology of breast tumors. J Japanese Soc Clin Cytol 37:43–49
9. http://www.hospitex.it/cyto_method.php
10. Fernàndez C, Minton AP (2008) Automated measurement of the static light scattering of macromolecular solutions over a broad range of concentrations. Anal Biochem 381:254–257
11. NSG-SELFOC collimator, http://www.nsgamerica.com/files/pdf/collimator.pdf

Chapter 38
Hybrid Ring-Resonator Optical Systems for Nanoparticle Detection and Biosensing Applications

C. Ciminelli, C.M. Campanella, and M.N. Armenise

In this work we investigate a method to detect and size nanoparticles interacting with an optical microresonator. Quantum electrodynamics principles, i.e. quantum harmonic oscillators properties, have been employed to model the interaction between the planar resonator and a spherical nanoparticle. Preliminary results show that the proposed method and the designed resonator are well suitable for nanoparticles sizing giving a 10% error in 30 nm radius nanoparticle characterization.

1 Introduction

Technological advances, due to the development of nanotechnology, give rise to fine and ultrafine particles production. The latter, also called nanoparticles, have at least one dimension equal or even smaller than 0.1 μm. Due to the small size, they exhibit quantum effects and have a higher ability to damage living organisms than microparticles. Nanoparticles can be generated via several synthetic methods based on gas, liquid or solid phase approaches and exist with great chemical and morphological diversity.

Different approaches have been employed in environmental parameters measurements in order to detect and size nanoparticles dissolved in a fluid. Optical resonant cavity structures supporting whispering gallery modes (WGMs), i.e. modes propagating inside a circular guide via the phenomenon of the total internal reflection and showing really high value of quality factor Q due to minimal reflection losses, seem to be the most promising configurations for nanoparticles detection, due to their high sensitivity and resolution, parameters both strictly related to Q.

C. Ciminelli (✉) • C.M. Campanella • M.N. Armenise
Dipartimento di Elettrotecnica ed Elettronica, Politecnico di Bari, Bari, Italy
e-mail: ciminelli.caterina@gmail.com

A. D'Amico et al. (eds.), *Sensors and Microsystems: AISEM 2011 Proceedings*, Lecture Notes in Electrical Engineering 109, DOI 10.1007/978-1-4614-0935-9_38,

2 Modelling

Two different microresonant cavities, i.e. a ring resonator and a hybrid resonator which is an annulus with optical properties similar to the ones of a disk, each of them coupled to two different straight waveguides, have been considered during this study.

First, microcavity design, whose electromagnetic characterization is based on [1], has been optimized in order to achieve a high Q-factor by using a 2D FDTD commercial software package. With a radius small enough as 5 μm and a gap of 300 nm we have obtained a Q value roughly of 6×10^4 for the hybrid structure, that is twice the Q-value of the ring configuration, and the cavity optical behavior has been investigated in presence of a perturbation occurring along the light path. This perturbation can be due to either the presence of structural imperfections associated to the fabrication process or to the presence of impurities (ultra-fine dusts or nanoparticles) or to both sources. Since a portion of light is backscattered by this kind of defects along the light path, the coupling between degenerate WGMs propagating in clockwise (CW) and counter-clockwise (CCW) directions inside the cavity occurs. We exploit, according to [2,3], quantum electrodynamics (QED) principles in order to derive the interaction equations, where the nanoparticle operates as a Rayleigh scattering source.

2.1 *Master Equations*

The system under investigation is well approximated by a coupled harmonic oscillator system composed by two WGMs (CW and CCW) with degenerate resonant frequency and by n reservoir modes, i.e. all those matter oscillators, such as phonons, additional photon modes, etc., confined in the cavity and interested by the scattering and the damping of cavity eigenmode occurring when light is perturbed along its path (see Fig. 38.1).

The evolution of this coupled harmonic oscillator system can be described by a state equation that, properly manipulated, gives rise to two uncoupled equations defining the symmetric and the asymmetric standing wave modes SWMs amplitude respectively, appearing in the resonator transmission spectrum as a doublet, whose expression is:

$$A_+ = \frac{ku}{t + \Gamma + i(\Delta\omega - 2g)}$$
$$A_- = \frac{ku}{t + i\Delta\omega} \tag{38.1}$$

As from Eq. 38.1, the symmetric mode dip is broadened of Γ and shifted of 2 g with respect to the asymmetric mode, where Γ is the damping rate due to Rayleigh

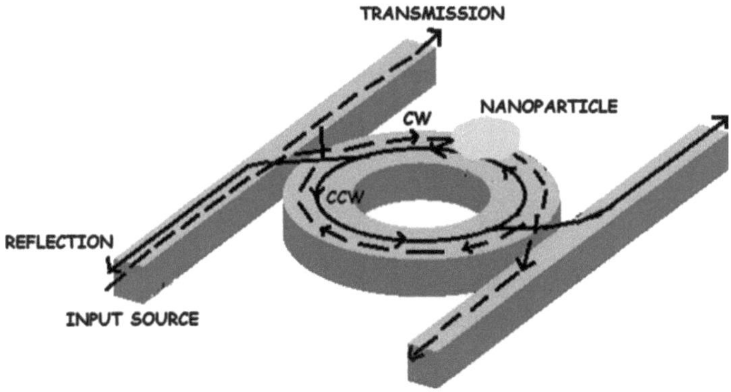

Fig. 38.1 Geometrical structure of a planar optical microresonator coupled to two straight waveguides and interacting with a nanoparticle (not in scale)

scattering, g is the coupling coefficient between symmetric and asymmetric mode, u is the input source amplitude, t is a parameter related to photon time-life inside the cavity, k is a parameter related to the extrinsic quality factor, and $\Delta\omega$ is the laser detuning frequency. Assuming the dipole approximation and by evaluating the distance between the two doublet dips and their linewidth, according to [2], we are able to derive the radius r of a spherical nanoparticle:

$$r = \frac{\lambda}{2\pi n_{bck}} \sqrt[3]{\frac{3}{4}\frac{\Gamma}{g}\frac{(n_p^2 + 2n_{bck}^2)}{(n_p^2 - n_{bck}^2)}} \tag{38.2}$$

where n_p is the nanoparticle refractive index, n_{bck} is the background refractive index and λ is the resonance wavelength before splitting.

3 Results

Simulations have been carried out for polystyrene (PS) and carbon tetrachloride (CCl_4) nanoparticles, with tested sizes in the range 30–100 nm, and showed that the CW and CCW coupling is affected by a modal splitting, a band broadening and a resonance shift rising, all occurring when nanoparticle dimensions increase. For a fixed nanoparticle refractive index, the perturbation of light travelling inside the resonator is proportional to nanoparticle radius. Thus, a consistent shift of the resonance wavelength, as in Fig. 38.2, occurs for both symmetric and asymmetric mode.

By using Eq. 38.2, we evaluated nanoparticles radius. Results show that the hybrid resonator is more able to detect smaller size (i.e. R = 30 nm and R = 50 nm) nanoparticles than the ring configuration, due to its higher sensitivity and quality factor (see Fig. 38.3). Even if the Q value is smaller than the one of the sphere [2], the hybrid resonator shows a good ability to sense the nanoparticle presence with an

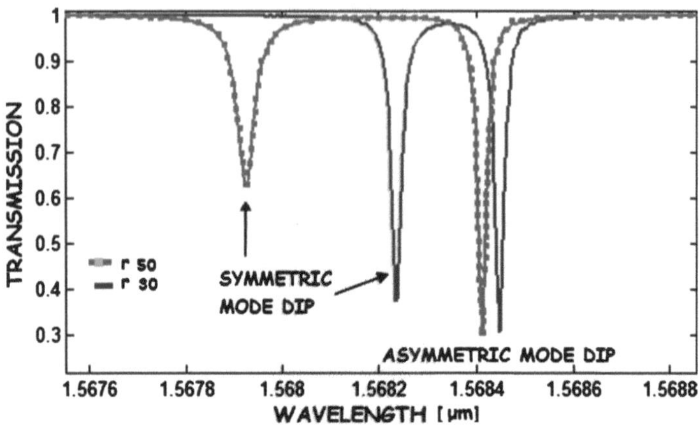

Fig. 38.2 Transmission spectra of the hybrid resonator when interacting separately with a PS nanoparticle of radius R = 30 nm and then R = 50 nm. Modal splitting and shift of WGM are clear

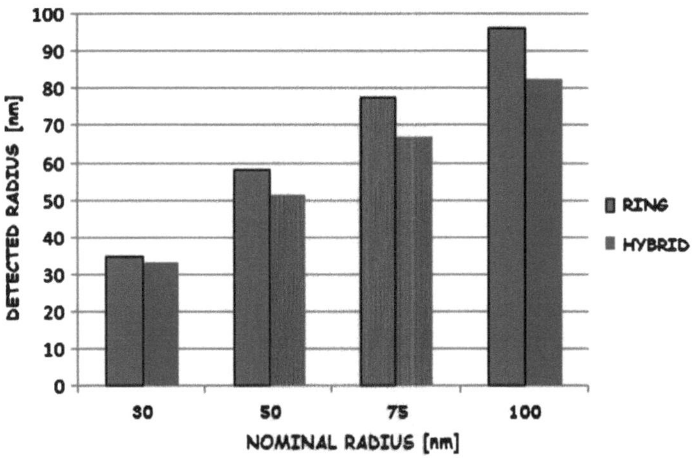

Fig. 38.3 Detected radius of a PS nanoparticle interacting separately with the ring and the hybrid microresonator

error within 10% in small nanoparticle radius (R = 30 nm) evaluation and it is more flexible from a technological point of view.

4 Conclusions and Future Outlook

Here we supposed to operate in an ideal condition, where a single nanoparticle-resonator interaction occurs, nanoparticle refractive index is known and nanoparticle shape is perfectly spherical. Work is in progress to develop a technique not only for identifying also the refractive index of the nanoparticle through the modeling of resonator transmission spectra, but also for modeling the interaction between a cloud of nanoparticles and the microresonator.

References

1. Ciminelli C, Campanella CM, and Armenise MN (2010) Simulation and fabrication of a new photonic biosensor. Proceedings of ICTON 2010, Munich, 27 June–1 July 2010
2. Zhu J, Ozdemir S.K., Xiao YF, Lin Li, Lina He, Da-Ren Chen, Lan Yang (2010) On-chip single nanoparticle detection and sizing by mode splitting in an ultrahigh-Q microresonator. Nat Photonics 4:46–49
3. Mazzei A, Götzinger S, de Menedes LS, Zumofen G, Benson O, Sandoghdar V (2007) Controlled coupling of counter-propagating whispering-gallery-modes by Rayleigh scatterer. Phys Rev Lett 99 pp. 173603-1–173603-4

Chapter 39
High-Order One-Dimensional Silicon Photonic Crystals with a Reflectivity Notch at $\lambda = 1.55\ \mu m$

S. Surdo, L.M. Strambini, G. Barillaro, F. Carpignano, and S. Merlo

In this work vertical, high aspect-ratio one-dimensional photonic crystals (1D-PhCs) featuring a reflectivity notch at $\lambda = 1.55\ \mu m$ were designed and fabricated by silicon electrochemical micromachining. Optical characterization of as-fabricated 1D-PhC structures was performed by acquiring reflectivity spectra, in the infrared region, on different locations of the devices, at normal incidence. Experimental data are in good agreement with theoretical reflectivity spectra, which were calculated by taking into account non-idealities of both the 1D-PhC structure and the measuring setup.

1 Introduction

In the last 10 years, photonic crystal (PhC) structures have attracted a lot of interest for optical communications and biosensing applications [1]. PhCs are natural or artificial materials characterized by spatial periodic variations of the dielectric constant. These materials exhibit photonic bandgaps (PBGs), i.e. wavelength intervals in which the propagation of the electromagnetic radiation is forbidden.

In our previous work, we reported design and fabrication of vertical, silicon/air one-dimensional photonic crystals (1D-PhCs) featuring high-order PBGs in the near infrared region [2]. However, for biosensing and optofluidic applications, high-order PhCs having a reflectivity minimum with high quality factor Q (notch), in place of a reflectivity maximum (bandgap), around the wavelength $\lambda = 1.55\ \mu m$ would be more appealing, the higher the Q factor value the higher

S. Surdo • L.M. Strambini • G. Barillaro (✉)
Dipartimento di Ingegneria dell'Informazione: Elettronica, Informatica, Telecomunicazioni, Università di Pisa, Pisa, Italy
e-mail: g.barillaro@iet.unipi.it

F. Carpignano • S. Merlo
Dipartimento di Elettronica, Università degli Studi di Pavia, Pavia, Italy

A. D'Amico et al. (eds.), *Sensors and Microsystems: AISEM 2011 Proceedings*, Lecture Notes in Electrical Engineering 109, DOI 10.1007/978-1-4614-0935-9_39,

the sensitivity of the structure [3]. In this work we report the design, fabrication, and optical characterization of 1D-PhCs featuring a reflectivity minimum with $Q \approx 3{,}000$ around the wavelength $\lambda = 1.55$ μm.

2 Design, Fabrication and Optical Characterization of 1D-PhCs with a Reflectivity Notch at $\lambda = 1.55$ μm

1D-PhCs featuring a reflectivity notch at 1.55 μm were designed according to the so-called hybrid quarter-wavelength stack. The structures consist in periodic arrays of silicon layers with thickness d_{Si} and air gaps of width d_{Air} satisfying the conditions $d_{Si} = M\lambda/4n_{Si}$ and $d_{Air} = N\lambda/4$, where M and N are even integer independent parameters, n_{Si} is the silicon refractive index at λ (whereas $n_{Air} = 1$). Design parameters were the silicon refractive index value $n_{Si} = 3.48$, the spatial period $p = 8$ μm and the porosity of the structure $P = d_{Air}/(d_{Air} + d_{Si}) = 0.554$, which corresponds to silicon walls of 3.57 μm (M = 32) separated by 4.43-μm-thick air gaps (N = 12). As-designed 1D-PhCs were fabricated by using the silicon electrochemical micromachining (ECM) technology, as detailed in Ref. [4]. A Scanning Electron Microscopy (SEM) section-view of a typical 1D-PhC fabricated by ECM technology is reported in Fig. 39.1. The high quality of the microfabricated structures, in terms of uniformity (X, Y, and Z direction) and low surface (X-Y plane) roughness, is clearly highlighted.

Characterization of as-fabricated 1D-PhCs was performed by scanning the device on the X-Y plane with high spatial-resolution steps and recording spectral reflectivity at normal incidence (in the wavelength range 1.0–1.7 μm) on different locations of the X-Y plane, using the optical setup detailed in [2]. Standard telecommunication single-mode fibers (SMR) with tapered lensed termination were used for back-reflection measurements. The device was exposed to white light radiation by means of a single-mode fiber-optic 50% coupler, which also carried the reflected light back to the optical spectrum analyzer (OSA).

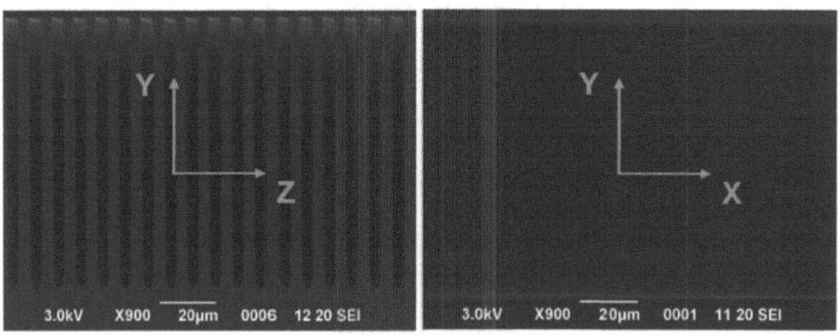

Fig. 39.1 SEM section-views of a typical 1D-PhC fabricated by silicon ECM technology

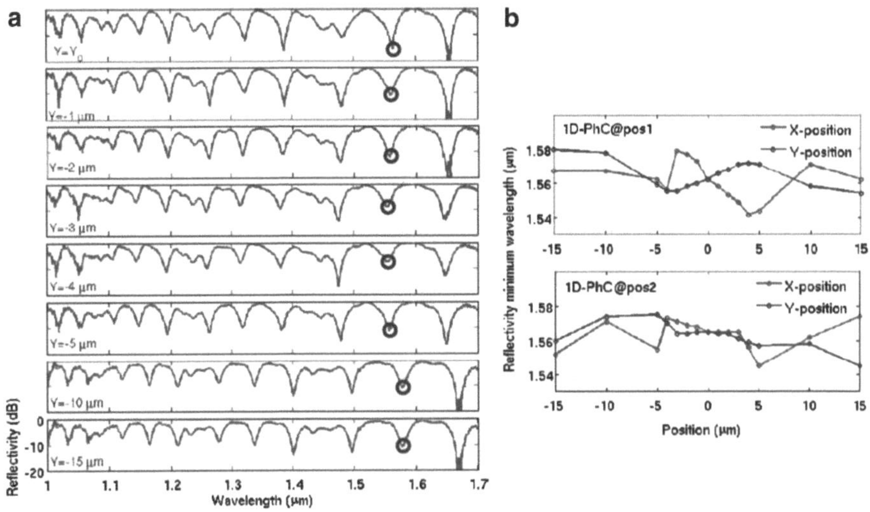

Fig. 39.2 (**a**) Experimental normalized reflectivity spectra collected by Y-scan from the reference position down to −15 μm. (**b**) Wavelength position of the reflectivity notch around 1.55 μm as a function of the X-Y position for two different starting locations on the same sample

Figure 39.2a shows the normalized reflectivity spectra recorded along the Y-axis from the reference position (x_0, y_0) down to −15 μm. All the recorded reflectivity spectra are characterized by a reflectivity notch around 1.55 μm, as designed. Experimental results show a good reproducibility of the reflectivity spectrum line-shape, although a small shift of the spectrum to shorter or longer wavelengths, and in turn of the reflectivity notch, as function of the Y position is also visible. Figure 39.2b shows that the maximum variation of the reflectivity notch wave-length-position is ±20 nm around 1.56 μm in a displacement range of ±15 μm around two different reference positions on the X-Y plane.

Numerical calculations were performed to compare experimental reflectivity spectra with the theoretical reflectivity spectra, for each measurement position. Theoretical reflectivity spectra were calculated by using the characteristic matrix method, modified to take into account non-idealities of both the 1D-PhC structure (i.e. surface roughness) and the measurement setup (i.e. OSA resolution bandwidth – RB), as detailed in [2].

Figure 39.3a shows an experimental reflectivity spectrum compared with the theoretical spectrum, the latter calculated by taking into account the limited RB (RB = 10 nm) used in the measurements. The good agreement between measured and calculated reflectivity spectrum is clearly highlighted. In order to obtain an accurate estimation of the quality factor of the reflectivity notch around 1.55 μm a portion of the reflectivity spectrum around the notch position was recorded using RB = 0.1 nm (Fig. 39.3b). In this case, an amplified spontaneous emission (ASE) source was employed to ensure good signal-to-noise ratio. The quality factor of the

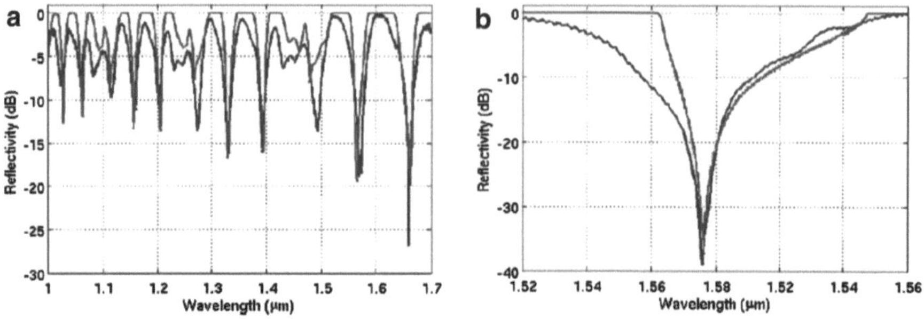

Fig. 39.3 Comparison between measured (*black trace, blue online*) and calculated (*gray trace, violet online*) reflectivity spectra collected on a high-order 1D-PhC: (**a**) RB = 10 nm. (**b**) RB = 0.1 nm

reflectivity minimum was estimated to be about 3,000, in agreement with theoretical predictions. It is worthy of being mentioned that such a Q value is rather high for 1D-PhCs not yet exploiting specific resonant-cavity effects.

3 Conclusions

In this work, 1D-PhC devices featuring a high quality factor reflectivity notch ($Q \approx 3{,}000$) around the wavelength $\lambda = 1.55$ μm were designed, fabricated and optically characterized. Reflectivity spectra acquired on different locations on the 1D-PhC device highlight a maximum shift of the notch position of ±20 nm over a range of ±15 μm around a reference location. Further work will be devoted to investigate possible correlation between fabrication-induced errors and notch position shift.

Acknowledgments This research was partially supported by PRIN-MIUR and Cariplo Fundation. Currently, F. Carpignano and S. Surdo hold a fellowship funded by Fondazione Alma Mater Ticinensis, Pavia, Italy.

References

1. Soref R (2010) Silicon photonics: a review of recent literature. Silicon 2(1):1–6
2. Barillaro G, Strambini LM, Annovazzi-Lodi V, Merlo S (2009) Optical characterization of high-order 1-D silicon photonic crystals. J Sel Top Quantum Electron 15:1359–1367
3. Barillaro G, Merlo S, Strambini LM (2009) Optical characterization of alcohol-infiltrated one-dimensional silicon photonic crystals. Opt Lett 34:1912–1914
4. Barillaro G, Diligenti A, Benedetti M, Merlo S (2006) Silicon micromachined periodic structures for optical applications at $\lambda = 1.55$ μm. Appl Phys Lett 89(15):151110-1–151110-3

Chapter 40
Distributed Strain and Temperature Sensing at CM-Scale Spatial Resolution by BOFDA

Romeo Bernini, Aldo Minardo, and Luigi Zeni

In this work we demonstrate high spatial (≈cm) and spectral (≈MHz) resolution Brillouin sensing by use of Brillouin Optical Frequency-Domain analysis (BOFDA). The employed set-up was capable to resolve a 3 cm spot perturbation at the far end of a 210 m fiber with a frequency resolution of 0.84 MHz. The method has potentialities for mm-scale spatial resolution and long sensing ranges.

1 Introduction

Distributed fiber sensors based on Brillouin scattering have been studied for over two decades, and several interesting schemes have been developed in the form of reflectometry or analysis for the measurement of local Brillouin frequency (ν_B). Time-domain approach has several benefits, among which a straightforward interpretation of the retrieved traces, as any time instant on the acquired trace corresponds to a well-defined section along the fiber through the time-of-flight of the pulse. However, the standard time-domain scheme suffers from a fundamental limitation related to the temporal width of the interrogating pulses. When the duration of these pulses is reduced below ≈10 ns (corresponding to a spatial resolution ≈ 1 m), the amplification of the Stokes beam obeys to a broadened and weaker gain spectral curve [1]. This results in a low signal-to-noise (S/N) ratio and low accuracy in the retrieval of ν_B. Different approaches have been proposed to overcome this limitation. The more relevant methods are those based on: (1) dark pulses [2]; (2) differential pulse-width pairs [3]; and (3) dynamic

R. Bernini (✉)
Istituto per il Rilevamento Elettromagnetico dell'Ambiente – Consiglio Nazionale delle Ricerche, Naples, Italy
e-mail: bernini.r@irea.cnr.it

A. Minardo • L. Zeni
Dipartimento di Ingegneria per l'Informazione, Seconda Università di Napoli, Aversa, Italy

A. D'Amico et al. (eds.), *Sensors and Microsystems: AISEM 2011 Proceedings*, Lecture Notes in Electrical Engineering 109, DOI 10.1007/978-1-4614-0935-9_40,

Brillouin gratings [4]. By these methods cm- or even mm-scale spatial resolution has been demonstrated. All the mentioned techniques operate in the time-domain, i.e. rely on some form of pulsing of the pump beam. As an alternative, a frequency-domain method (BOFDA) is applicable to obtain enhanced spatial resolution in Brillouin-based distributed sensors [5, 6]. In this approach, the pump beam is not pulsed; rather its intensity is sinusoidally modulated. The Brillouin gain curve is retrieved by detecting the intensity modulation induced on the Stokes beam over a range of modulation frequencies. The main difference with time-domain approaches stems from the fact that in BOFDA systems the optical carriers associated to the Stokes and pump beams form an intense, stationary acoustic wave, acting as a pre-trigger for measuring the interaction between the Stokes beam and the ac pump component. As the stationary acoustic wave is much more intense than its ac counterpart, the Brillouin gain curve keeps its natural linewidth ($\approx$ 30 MHz in standard silica fibers), allowing to determine ν_B with high accuracy even if operating in the high-spatial resolution regime [6].

The paper presents a successful demonstration of BOFDA distributed measurements at cm-scale spatial resolution, with an improvement of more than one order of magnitude with respect to previous BOFDA demonstrations.

2 Theoretical Description

In the BOFDA scheme a cw probe wave is launched at z = L, while an intensity-modulated pump wave is launched at z = 0. The probe wave is phase- and amplitude- modulated while it propagates down the fiber due to SBS interaction with the pump. A complex baseband transfer function, defined as the ratio between the ac components of input pump and output probe, is acquired over a range of modulation frequencies, and for different values of the pump-probe frequency shift.

Positional information can be achieved by inverse Fourier transforming the acquired transfer functions. In a linear approximation, the spatial resolution can be calculated as [5]:

$$W = c/2n\left(f_{m,\max} - f_{m,\min}\right) \quad (40.1)$$

where $f_{m,max}$ and $f_{m.min}$ denote the maximum and minimum modulation frequency, respectively. The maximum fiber length that can be investigated is related to the minimum modulation frequency (we assume that $f_{m,min}$ also corresponds to the modulation frequency step) [5]:

$$L_{\max} = c/2nf_{m,\min} \quad (40.2)$$

In measurements involving very long ranges (km or tens of km), the maximum sensing length may be rather limited by pump depletion occurring along the whole fiber length. Pump depletion can be controlled by employing low levels of probe power, or by injecting both Stokes and anti-Stokes waves at the $z = L$ input [7]. By this technique, BOFDA measurements up to 5 km have been demonstrated.

3 Experimental Results

The BOFDA experimental configuration is depicted in Fig. 40.1. A 1,551 nm 40 mW distributed-feedback laser diode is used as the light source.

The polarized output of the laser diode is split into two distinct channels to allow both the pump and probe waves to be derived from the same optical source. The signal is generated in the upper channel, where an integrated electro-optic intensity modulator (IM1), driven by a radio frequency synthesizer in a suppressed carrier configuration, creates sidebands, one of which (the upper one) is eliminated by a narrowband (≈10 GHz) fiber Bragg grating (FBG) filter. In the lower channel a second electro-optic intensity modulator (IM2) driven by the RF output of the vector network analyzer is used to modulate the pump beam across a range of modulation frequencies. Both pump and probe waves are amplified before being injected in the sensing fiber. A polarization scrambler is employed to minimize the gain fluctuations due to changes in the state of polarization of pump along the sensing fiber. A wideband (3.5 GHz) photodetector with a transimpedence gain of 1.100 V/W is used to convert the received optical signal in an RF signal, which is then demodulated by the network analyzer. In the experiments, the maximum modulation frequency was 3.5 GHz (≈29 mm spatial resolution) due to the cutoff frequency of the photodetector. Furthermore, the network analyzer was set to operate with an IF bandwidth of 10 kHz and with a number of averages equal to 10.

As a first test case, we constructed a ≈9 m SMF sample consisting of nine cascaded fiber portions of type A ($\nu_B \approx$ 10,670 MHz) or type B ($\nu_B \approx$ 10,720 MHz). The four portions of fiber B had a length of 20, 10, 10 and 3 cm. Figure 40.2 reveals

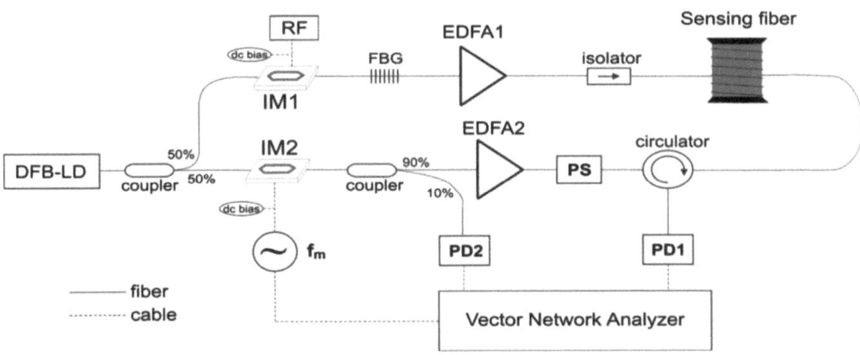

Fig. 40.1 Experimental setup for BOFDA measurements. DFB-LD: diode laser; EDFA: erbium-doped fiber amplifier. PS: polarization scrambler; IM: intensity modulator; PD: photodetector

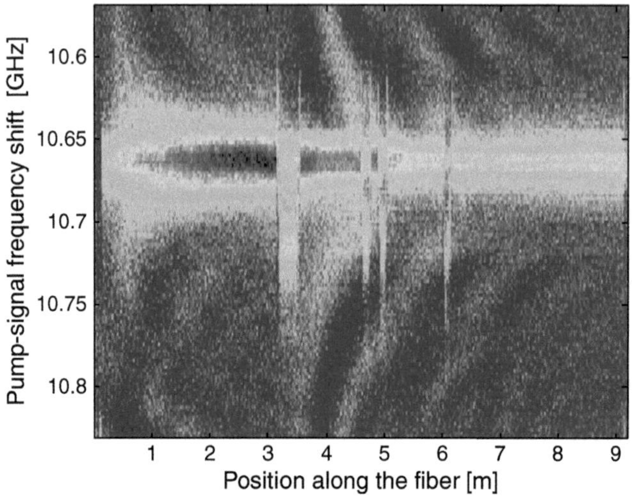

Fig. 40.2 Measured BGS distribution along the prepared SMF sample

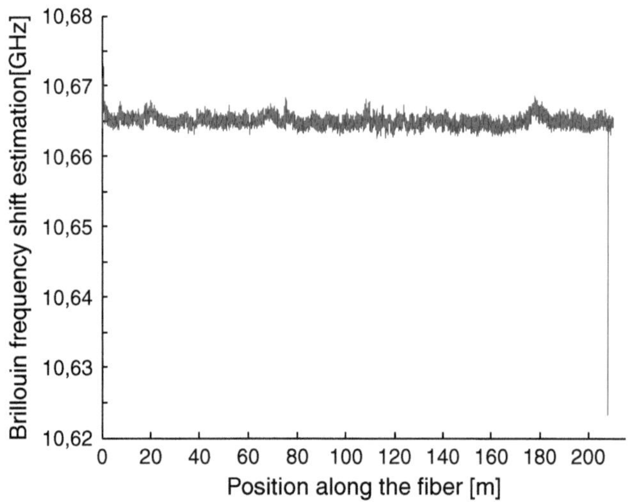

Fig. 40.3 Distribution of the Brillouin frequency shift along the 210 fiber strand with a 3 cm spot at the far end

that all four spots can be clearly distinguished from the inverse Fourier transformed acquired traces. Note that the off-resonance oscillations appearing in the measurements arise from the interaction between the acoustic wave ac component and the pump wave stationary component [6].

As a further example, we performed a measurement by placing a 3 cm piece of fiber C ($\nu_B \approx$ 10,620 MHz) at the far end of a ≈210 m strand of fiber A. This example was chosen aiming to demonstrate that high-resolution capabilities can be also achieved with longer fibers. In Fig. 40.3 we report the overall distribution of the ν_B as obtained by Lorentzian peak-searching the acquired BGS. The error on ν_B determination along the 3 cm spot is $\approx$ 3 MHz. The standard deviation of the retrieved ν_B profile along fiber A is 0.84 MHz.

4 Conclusions

Distributed sensing at cm-scale spatial resolution was experimentally demonstrated by applying a BOFDA technique. The method has several advantages when compared to time-domain schemes including high spectral accuracy, high signal-to-noise ratio and mm-scale spatial resolution capabilities. At the present, the main limit is the measurement time, that becomes especially long in case of high-resolution measurements over long ranges. Work is in progress in order to increase the measurement speed to such extent that measurements at cm-scale spatial resolution over tens-of-kilometer fiber lengths are possible in a reasonable time.

References

1. Lecoeuche V, Webb DJ, Pannell CN, Jackson DA (2000) Transient response in high-resolution Brillouin-based distributed sensing using probe pulses shorter than the acoustic relaxation time. Opt Lett 25:156–158
2. Brown AW, Colpitts BG, Brown K (2007) Dark-pulse Brillouin optical time-domain sensor with 20-mm spatial resolution. J Lightwave Technol 25:381–386
3. Li W, Bao X, Li Y, Chen L (2008) Differential pulse-width pair BOTDA for high spatial resolution sensing. Opt Express 16:21616–21625
4. Song KY, Zou W, He Z, Hotate K (2009) Optical time-domain measurement of Brillouin dynamic grating spectrum in a polarization-maintaining fiber. Opt Lett 34:1381–1383
5. Garus D, Krebber K, Schliep F, Gogolla T (1996) Distributed sensing technique based on Brillouin optical-fiber frequency-domain analysis. Opt Lett 21:1402–1404
6. Minardo A, Testa G, Zeni L, Bernini R (2010) Theoretical and experimental analysis of Brillouin scattering in single mode optical fiber excited by an intensity- and phase-modulated pump. J Lightwave Technol 28(2):193–200
7. Minardo A, Bernini R, Zeni L (2009) A simple technique for reducing pump depletion in long-range distributed Brillouin fiber sensors. IEEE Sensors J 9(6):633–634

Chapter 41
Cascaded LPG and FBG Integrated in a Miniaturized Flow Cell for Compensated Refractometric Measurement

Francesco Chiavaioli, Marco Mugnaini, Cosimo Trono, Francesco Baldini, and Massimo Brenci

The present paper describes the design and characterization of a thermo-stabilized flow cell for accurate refractive index (RI) measurements using a long period grating (LPG)-based sensor and a methodology to measure and correct all the LPG cross-sensitivity. A fiber Bragg grating (FBG) written on the same fiber in series with the LPG and an accurate measuring system of the temperature are used in order to minimize the interferences coming from temperature and strain changes. The experimental results show that the proposed system provides good performance as far as the RI sensitivity and resolution are concerned. The maximum RI sensitivity and resolution are around 1.455 RIU and the values are 3,120 nm RIU^{-1} and 2×10^{-5} RIU, respectively.

1 Introduction

Thanks to the advantages offered by the optical fiber sensors (OFS), refractive index (RI) OFS have been studied since many years and also used in the area of optical biochemical sensing. A well-known method is based on LPGs, in which the periodic modulation of the fiber core RI sets out the coupling between the fundamental core mode and the cladding modes that, by verifying the phase-matching conditions, generates a series of attenuation bands in the transmission spectrum of the fiber. These bands are centered at discrete wavelengths, i.e. resonance wavelengths, and are sensitive to changes in the external RI [1].

F. Chiavaioli (✉) • M. Mugnaini
Department of Information Engineering, University of Siena, Siena, Italy
e-mail: chiavaioli@dii.unisi.it

C. Trono • F. Baldini • M. Brenci
Institute of Applied Physics "Nello Carrara", National Research Council of Italy, Sesto Fiorentino (FI), Italy

A. D'Amico et al. (eds.), *Sensors and Microsystems: AISEM 2011 Proceedings*, Lecture Notes in Electrical Engineering 109, DOI 10.1007/978-1-4614-0935-9_41,

The present paper describes the design and characterization of a LPG-based refractometer integrated in a thermo-stabilized low-volume flow cell for reliable measurement in liquid samples. An ad hoc methodology has been developed for measuring and then correcting the cross-sensitivities coming from strain and temperature changes: using the theoretical expressions [2] and the experimentally determined coefficients of the cross-sensitivities, the shift of the LPG resonance wavelength only due to changes in the external RI is achieved, measuring both the temperature and the resonance wavelength of an FBG written in cascade along the same fiber.

2 Experimental Setup

2.1 Flow Cell

The cross-section of the flow cell is sketched in Fig. 41.1. The flow cell is made up of two pieces: the upper part is a 4 mm thick PMMA transparent layer for the visual examination of the flow cell and the bottom one is a 6 mm thick aluminum layer, positioned in thermal contact with a thermo electric cooler (TEC). The flow cell is 80 mm long, 15 mm wide and 10 mm high. The flow channel is 50 mm long with a cross-section of 1 mm^2, corresponding to a total volume of 50 μL. A specially shaped parafilm® sheet is inserted between the two bars for ensuring the waterproofing of the flow cell. The inlet and outlet of the flow cell consists of two stainless-steel tubes inserted on the PMMA layer. A thermocouple connected with a thermometric measuring unit (Lutron TM-917) measures the temperature of the flow cell which is acquired every 2 s by a PC with a resolution of ±0.01°C. It is worth noting that the real-time monitoring of the temperature is needed to control potential drifts induced by ambient temperature changes and to make allowances for the effects of pumping fluids at different temperatures within the flow cell.

2.2 Manufacturing of the Gratings

The FBG is fabricated with an Excimer KrF laser (Lambda Physics Compex) by exposing a B-Ge co-doped optical fiber (Fibercore PS1250/1500) through a

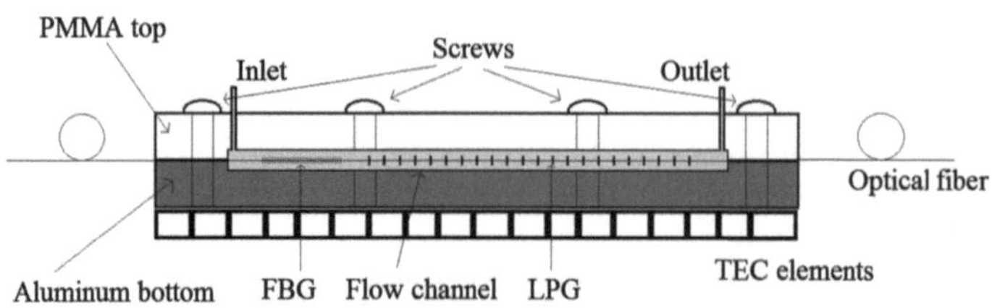

Fig. 41.1 Cross-section of the flow cell

rectangular phase mask (1,059.9 nm phase mask pitch) [3]. Their characteristics are: 1,534.2 nm resonance wavelength (λ_{res}), 90% reflectivity, 1 cm long and about 0.26 nm of full width at half maximum (FWHM). LPGs are inscribed by a point-to-point technique with the same laser but using a different optical setup: the fiber, fixed on a motorized translation stage (Burleigh 6000), is irradiated by the laser spot which is shaped and focused by means of a cylindrical lens. A PC controls both the translation stage and the laser action with an ad hoc developed NI CVI program and a commercial digital board (Eagle μ-DAQ USB-96C), in order to choose both the grating pitch and the number of laser shots. The LPGs characteristics are: 615 μm grating pitch, 1,567 nm λ_{res}, −15 dB transmission loss at the λ_{res}, 2.46 cm long and about 6.5 nm of FWHM. It is worth noting that the spectral distance of the attenuation bands of the two gratings is about 30 nm, which is sufficient to avoid the interference of the two bands by supposing a blue shift of the LPG λ_{res} up to about 20 nm.

2.3 *Interrogation Technique*

The whole system is made up of a series of elements described as follows: the optical source is a broadband superluminescent diode (INPHENIX IPSDD1503) and the transmission spectrum of the two gratings is acquired by means of a commercial optical spectrum analyzer (OSA, Anritsu MS9030A and MS9701B) with 0.1 nm optical resolution. The OSA is then connected to the PC and is controlled by means of an ad hoc developed acquisition program.

2.4 *Fluidic System and Chemicals*

The flow cell is connected to a peristaltic pump (Gilson MINIPULS 3) by means of PVC tubing (1.02 mm internal diameter), which allows to pump the proper solution into the flow cell. The used flow rate is 500 μL min^{-1}. The RI of the solutions (mixtures of glycerol and water in different volumetric ratios) is varied from 1.334 (distilled water) up to 1.467 (last test solution) and is measured by means of a commercial hand-held refractometer Atago R5000 (@ 23°C).

3 Results and Discussion

3.1 *Cross-Sensitivities Characterization*

The two gratings have been characterized in terms of strain and temperature. Figure 41.2 shows the plot of the resulting λ_{res} for the two gratings as a function

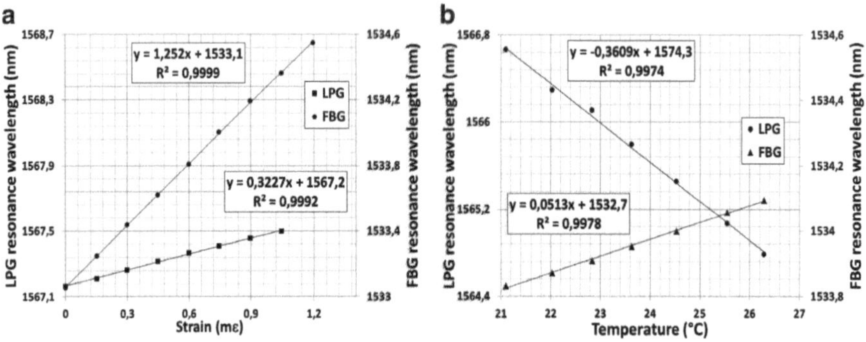

Fig. 41.2 Strain (**a**) and temperature (**b**) characterization of the two gratings

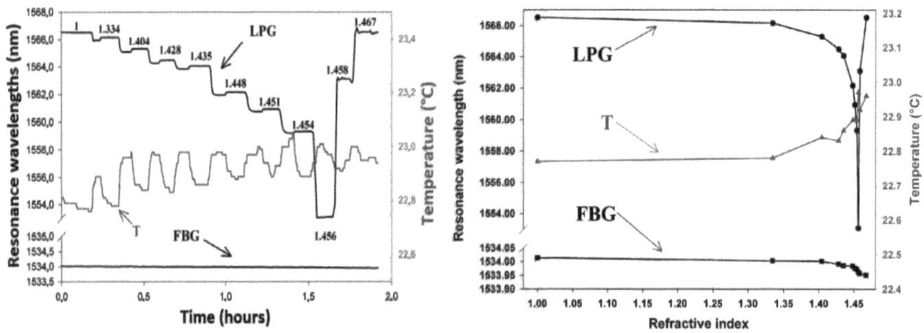

Fig. 41.3 Sensorgram (*left*) and response curve (*right*) of the proposed sensing system

of applied strain (41.2a) and temperature (41.2b), respectively. The results shown in Fig. 41.2 highlight the linear trend of both the gratings within their measurement range (0–1.2 mε and 21–27°C) and the obtained values of the four experimental coefficients are the following: $k_\varepsilon = 1.252$ pm $\mu\varepsilon^{-1}$ and $k_T = 51.3$ pm °C^{-1} for the FBG and $k_\varepsilon = 0.323$ pm $\mu\varepsilon^{-1}$ and $k_T = 361$ pm °C^{-1} for the LPG.

3.2 *Refractive Index Measurement*

In order to achieve the response curve of the proposed refractometer, different solutions of glycerol and water were pumped into the flow cell, locking the TEC at a constant temperature of about 23°C. Figure 41.3, on the left, shows the changes in λ_{res} values of the two gratings as a function of time (sensorgram), together with the temperature evolution measured by the thermocouple. As expected, the LPG λ_{res} shows a great blue shift when the RI of the solution is approaching the RI of the fiber cladding (1.455 RIU) [1, 4]. All these data are corrected by means of both the shift of the FBG resonance wavelength and the measured temperature. The refractometer response curve (Fig. 41.3, right) was obtained considering the plateaus in which the flow has been stopped, and estimating the mean value of both the λ_{res}. As mentioned above, the maximum sensitivity is achieved around 1.455 RIU and the calculated value is 3,120 nm RIU^{-1}. By defining the sensor resolution as three times the standard deviation on the previously-determined sensitivity, the maximum resolution is 2×10^{-5} RIU.

4 Conclusions

A LPG-based refractometer with its flow cell and its thermal stabilization section, together with the method for measuring and then correcting the LPG cross-sensitivities have been described. Manufacturing of the gratings and the interrogation technique have been presented too. The proposed sensing system makes it also possible to measure other parameters, such as strain and temperature. Experimental results prove that the system performance is good in terms of the RI sensitivity and resolution, with the best value of 3,120 nm RIU^{-1} and 2×10^{-5} RIU, respectively, obtained around 1.455 RIU.

References

1. James SW, Tatam RP (2003) Optical fibre long-period grating sensors: characteristics and application. Meas Sci Technol 14:R49–R61
2. Kersey AD, Davis MA, Patrick HJ, LeBlanc M, Koo KP, Askins CG, Putnam MA, Friebele EJ (1997) Fibre grating sensors. J Lightwave Technol 15:1442–1463
3. Hill KO, Meltz G (1997) Fibre Bragg grating technology fundamentals and overview. J Lightwave Technol 15:1263–1276
4. Patrick HJ, Kersey AD, Bucholtz F (1998) Analysis of the response of long period fiber gratings to external index of refraction. J Lightwave Technol 16:1606–1612

Chapter 42
An Investigation on the Double Nature of Photons

Pasquale Acquaro

The theory of special relativity excludes that a material body can move at the speed of light, but it does not exclude at all that light can be made of material points.

The Einstein's equation relating the mass of a particle to its velocity is used to decide if a photon has a mass or not. Herewith it is suggested that it should instead be used only to calculate the relativistic mass of a particle that, when it is accelerated, reaches the velocity v.

Photon mass can be argued from the Planck equation relating the energy and the frequency. Considering the expression of the Planck length it is possible to determine a relationship evidencing the mass of the photon.

In this paper, some experiments will be analyzed, that have confirmed the existence of a photon mass and the equations of the special relativity will be used, as they allow us to ascertain that also Einstein had considered the concept of photon mass as basis for his own equations.

Furthermore, some thought experiments illustrating both the acceleration of elementary particles and the determination of the radiation pressure will be illustrated. The latter will be calculated considering the photon as a material corpuscle that moves following the Newton's Laws of dynamics.

In addition, it will be shown that during scattering with material particles photon transfers its mass, its momentum and its wavelength to the particle, and that the wavelength perfectly complies with the de Broglie wavelength.

P. Acquaro (✉)
Via A. De Gasperi trav 20, 89900 Vibo Valentia, Italy

Campus Biomedico di Romavia Alvaro del Portillo, 21 00128 Rome (Italy)
e-mail: pasqualeacquaro@virgilio.it

A. D'Amico et al. (eds.), *Sensors and Microsystems: AISEM 2011 Proceedings*, Lecture Notes in Electrical Engineering 109, DOI 10.1007/978-1-4614-0935-9_42,

1 Introduction

The problem of the double nature of light as a corpuscule and wave, was first raised in the seventeenth century. During the Solvay Congress that took place in Como in 1927, and in which also Planck and Einstein took part, Bohr formulated the “complementarity principle”, according to which wave and particulate aspects of light should emerge in different contexts and should mutually exclude each other [1]. But is this true? A different reality may be found exploring the relativity equations; putting in evidence how the two aspects of light are closely related [1].

Einstein when formulated the special relativity was convinced that “light quanta” have a wave behavior, which is not, however, separated from the corpuscular behavior. Before him, in 1675, Newton [2] thought that light corpuscles received vibrations from the ether, by interfering with it.

In this paper I intend to show, inter alia, that photons have both the characteristics of a corpuscle with mass and those of a wave, and that their mass is directly proportional to the frequency of the wave they show.

2 Photon Mass

Let us consider the Planck length

$$L_P = \left[\frac{Gh}{c^3}\right]^{\frac{1}{2}} = 4.05 \cdot 10^{-33} cm \tag{42.1}$$

from this relation it is possible to derive the Planck constant, that substituted in the Planck energy relationship $E = h \cdot \upsilon$ gives:

$$E = \frac{L_P^2 c^3}{G} \upsilon \tag{42.2}$$

This can be rearranged as:

$$E = \left[\frac{L_P^2 c}{G} \upsilon\right] \cdot c^2 \tag{42.3}$$

where the quantity in square bracket has evidently the dimension of a mass.

The above equations illustrate that through the Planck Length the energy of mass particle (E = mc^2) and photons ($E = h \cdot \upsilon$) coincide.

Equation 42.3 appears more significant than the Planck equation because it closely links the mass of the photon with its frequency and expresses three fundamental constants: L (Planck length), c (speed of light in vacuum), and G (gravitational constant).

As a consequence, it is possible to affirm that:

$$m = k \cdot \upsilon \text{ with k} = \frac{L_P^2 c}{G} = 7.372 \cdot 10^{-48} g \cdot s \tag{42.4}$$

The constant k can be derived from another expression obtained by the author as a consequence of an alternative model of Universe [3].

$$\left[\frac{m_p \cdot T(t)^2 \cdot V(t)^2 \cdot t_p \cdot d^3 \cdot G^2 \cdot L_p^3}{N \cdot T_p \cdot k_w \, c^5}\right]^{\frac{1}{2}} = \frac{L_P^2 c}{G} \tag{42.5}$$

where: m_p = Planck mass = 5.46×10^{-5} g; T(t) = temperature of the Universe at the cosmic time t; V(t) = volume of the Universe at the cosmic time t; t_p = Planck time = 1.35×10^{-43} s; d = nuclear density = $2{,}918 \times 10^{14}$ g/cm^3; G = gravitational constant = 6.67×10^{-8} [cm^3/(g × sec^2)]; L_p = Planck length = 4.05×10^{-33} cm; N = ratio between the Planck density and the nuclear density = 2.815 × 10^{78}; T_p = Planck temperature = 7.168×10^{31} K; Kw = Wien's constant = 0.29 cm K; c = speed of light in vacuum = 2.9979×10^{10} cm/s; h = Planck constant = 6.626×10^{-27} erg s; ν = frequency of the radiation in Hertz; V(today) = 1.384×10^{91} cm^3; T(today) = 2.725 K.

This constant, which is far more complex and articulated than that in Eq. 42.4 shows that massive photons had, and still have today, a very important role in the history of the Universe.

3 Connection with Special Relativity

The knowledge of the quantization of energy that Planck had theorized in order to explain the distribution of the black body radiations, led Einstein to understand that every energy quantum had to be associated with a "light quantum", which, according to him, was the equivalent of a massive photon.

By introducing these "light quanta", he theorized the mass increase that particles receive after their absorption, and was able to explain the photoelectric effect. But at that time, it was impossible to prove through experiments that light quanta had a mass. So Einstein was very cautious about affirming the particulate nature of light.

Only after the experiments carried out by Arthur Compton, W. Bothe and H. Geiger, which confirmed the validity of the relativistic laws of conservation of momentum, Einstein became fully convinced of the corpuscular nature of light quanta and completely understood the importance of his previsions. That happened about 25 years after the publication, in 1905, of the "electrodynamics of moving bodies", that is the publication of the theory of special relativity [4].

In this theory, the following equation shows that the light quanta theorized by Einstein has a mass.

$$m = \frac{m_0}{\left[1 - \left(\frac{v^2}{c^2}\right)\right]^{\frac{1}{2}}} \tag{42.6}$$

Inverting the previous equation we obtain:

$$v = c \cdot \left[1 - \left(\frac{m_0 c^2}{m_0 c^2 + E_0}\right)^2\right]^{\frac{1}{2}} \tag{42.7}$$

It shows that if we provide a particle at rest with a mass m_0 with radiant energy E_0, the particle moves and produces a velocity v.

Einstein considered that the radiation E_0 was made up of "point particles with energy hυ" which once absorbed by a body with a rest mass m_0, determine in it the "increase in its inertial mass by a quantity $m = E_0/c^2$" [5]. So we can conclude that the energy E_0, which determines the velocity v in the rest mass particle m_0, should also add to this particle the mass $m = E_0/c^2$.

For example, if an electron with a rest mass of 9.109×10^{-28} g being motionless absorbs a gamma photon with an energy of 1.9878×10^{-4} erg, it produces, according to Einstein's equation, a velocity of 2.9978×10^{10} cm/s, which is almost equal to the speed of light.

Equation 42.6 tells us that the electron accelerated up to the speed of 2.9978×10^{10} cm/s receives a mass increase of 221.176×10^{-27} g, which perfectly corresponds to that provided by the equation $m = E_0/c^2$.

This shows that the photon has determined both the mass increase and the velocity v in the electron. If we gave to the equation $E_0 = m c^2$ the meaning of energy, which is able to turn into mass, we should necessarily admit that the particle could not move, considered that it receives only mass (and not energy) from the photon.

But why, still today, do experts think that photons cannot have a mass? Because they improperly use Eq. 42.6 to support this thesis: This equation cannot be used for photons because they do not have a rest mass. If they had it, it should gradually grow with the increase in their velocity and should become infinite once the speed of light is reached. But we know very well that this cannot be true, considered that the speed of light is a constant value, confirmed by a universally-known physical principle.

4 Explaining the Radiation Pressure Using Newton Laws of Motion

It can be shown that a massive photon, unlike a photon without mass, produces a radiation pressure, which can be explained by the equations of Newton's laws of motion.

For example, if we put a photon with a frequency of 2.698×10^{11} Hz in a volume of 1 cm^3 (no matter its shape), it produces a pressure $P = n\,h\nu$ equal to 1.7909×10^{-15} dyn/cm^2. If the same photon is placed in a cube with a side of 1 cm, we know that it takes 3.333×10^{-11} s to horizontally move at the speed of light between two opposite walls.

At the moment of impact on the wall, it instantaneously stops and, since its velocity is annulled during the time between the two following impacts, it undergoes a deceleration of 9.009×10^{20} cm/s. If this photon didn't have a mass, it would be impossible to calculate its impact force.

Considered that its mass is $m = 7.372 \times 10^{-48} \times 2.698 \text{ x } 10^{11} = 1.988 \times 10^{-36}$ g, it determines on the wall of 1 square centimeter an impact force equal to 17.909×10^{-16} dyn and a pressure of 1.7909×10^{-15} dyn/cm^2, which is the same as that calculated by the equation $P = n\,h\nu$.

If we change the shape of the container, making it become a very lengthened prism, but keeping its volume of 1 cm^3 constant, between the two following impacts the photon covers a distance longer than the previous one. This entails a lower acceleration because, since the time between the two following impacts is extended, acceleration decreases. But despite this, the pressure is equal to 1.7909×10^{-15} dyn/cm^2, which is the same as that recorded in the cubic container. This happens because the wall the photon impacts on is far smaller than the one of the previous cube.

In short, it is the geometry of the container that solves the problem.

5 Connections with the De Broglie Wavelength

It can be proved that a photon transfers not only its mass and its momentum, but also its wavelength to a particle that includes it. That means that the De Broglie wavelength is a property, which is imported from the photon, and not a property of a moving mass, as it is believed [6].

For example, consider a gamma photon with an energy of 0.5705 erg and a frequency of 8.610×10^{25} Hz.

This photon has a mass:

$m = k\nu = 7.372 \times 10^{-48}\ \nu = 7.372 \text{ x } 10^{-48} \times 8.610 \times 10^{25} = 6.347292 \times 10^{-22}$ g (this paper)

$m = E/c^2 = 0.5705/(2.9979 \times 10^{10})^2 = 6.347292 \times 10^{-22}$ g (Einstein)

and a momentum:

$h\nu/c = 0.5705/2.9979 \times 10^{10} = 1.902998 \text{ x } 10^{-11}$ g $\times$ (cm/s) (Einstein)

$m\ c = 6.347292 \times 10^{-22} \times 2.9979 \times 10^{10} = 1.902854 \text{ x } 10^{-11}$ g $\times$ (cm/s) (this paper)

If this photon is absorbed by a proton at rest, whose mass is 1.67252×10^{-24} g, it transfers both its mass and its momentum to the proton. Hence the proton increases its mass by $636.40172 \times 10^{-24}$ g and produces the velocity $v = 2.9900 \times 10^{10}$ cm/s, which is very close to the speed of light.

The de Broglie equation informs us that this moving material particle is associated with a wavelength $\lambda = h/(m\ v) = 6.626 \times 10^{-27}/(638.074 \times 10^{-24} \times 2.9900 \text{ x } 10^{10}) = 3.473 \times 10^{-16}$ cm.

The photon we have considered has a frequency of 8.610×10^{25} Hz, to which corresponds a wavelength $\lambda = c/\nu = 2.9979 \times 10^{10}/8.610 \times 10^{25} = 3.482 \times 10^{-16}$ cm, which is identical to the de Broglie wavelength.

This shows that the photon has transferred both its mass and its wavelength to the particle, and that if it hadn't had them, it could not have transferred them to the particle. This may raise serious doubts about the validity of Bohr's complementarity principle.

6 Photons General Physical Law

It is easy to verify that any photon follows the following physical law recently formulated by the author:

$$\frac{\mathrm{M}_p \cdot \mathrm{T}_p}{\upsilon^2} = 7.131 \cdot 10^{-59} = \frac{L_p^2 k_w}{G} \tag{42.8}$$

This law has also important implications in Cosmology [3].

For example, it is known that the Universe at the Planck time (about 10^{-43} s after the Big Bang) was filled with individual energy particles equal to $4{,}907 \times 10^{16}$ erg, but until now, nobody has been able to affirm that these particles were photons. The above mentioned equation can give us a credible answer. If we associate the energy particles of 4.907×10^{16} erg with the frequency 7.405×10^{42} Hz ($\lambda = 4.05 \times 10^{-33}$ cm) and the mass of 5.46×10^{-5} g (Planck mass), and if we introduce these quantities in the equation above, we can obtain the corresponding temperature:

$$\mathrm{T}_p = 7.131 \cdot 10^{-59} \frac{\upsilon^2}{\mathrm{M}_p} \tag{42.9}$$

This temperature is equal to 7.164×10^{31} K and it coincides with the Planck temperature, which cosmologists attribute to the Universe just at the time in which the particles with the energy of $4{,}907 \times 10^{16}$ erg were present. The same temperature is also provided by Wien's law.

References

1. Bergia S, Einstein, Quanti e Relatività (1998) Una svolta nella fisica teorica. I grandi della scienza, anno I, n.6 (in Italian)
2. Guicciardini N, Newton (1998) Un filosofo della natura e il sistema del mondo. I Grandi della scienza, le scienze, anno i, n.2, aprile (in Italian)
3. Acquaro P (2008) Un microscopico buco nero all'origine dell'universo. Edizioni Monteleone (in Italian)
4. Einstein A (1905) Sull'elettrodinamica dei corpi in moto, Annalen der Physik, band 17
5. Einstein (2009) A Relatività: Esposizione divulgativa e Scritti Classici su Spazio, Geometria, Fisica, Fabbri Editori (in Italian)
6. Acquaro P (2001) Nuove equazioni per il calcolo delle proprietà degli elettroni accelerati [abstract]. XXXI congresso nazionale di chimica fisica, Padova, 19–23 giugno (in Italian)

Part V
Electronics and Technologies for Sensors

Chapter 43
Microfluidic System for Real Time PCR Sample Preparation

G. Barlocchi, F.F. Villa, and U. Mastromatteo

The possibility to make fast analysis of DNA in order to determine the presence of mutations, or the amount of gene expression (RNA), or other DNA based information using PCR, is becoming more and more important in the clinical diagnostics methodology. As a consequence, an increase of the performances of diagnostics tools requires the use of very simple, robust and automatic instrumental technologies. Concerning the DNA analysis systems, they exist nowadays several technologies able to satisfy such kind of requirements; among them, perhaps, the more competitive one and destined to a rapid expansion and integration inside lab tools, is the "REAL TIME PCR" [1]. This technique, as well known, allows to quantify the PCR reaction (direct measurement) while the reaction is ongoing. In the Real Time PCR Lab On Chip microsystem the most limiting factor for the usage expansion of such sophisticated diagnostic tool is due to the manual and time consuming procedure for the sample preparation. In order to overcome this limitation STMicroelectronics is working on a complete integrated system able to reduce to a minimum the number of steps for the sample preparation, integrating it directly on a fast operating Lab On Chip for Real Time PCR. Here in the following an automatic microfluidic system for RT-PCR is described.

1 DNA Sample Preparation Using Standard Spin Column-Based Purification Method

These days, most labs use commercial kits, which employ spin columns, for the isolation and purification of nucleic acids. The spin columns (Fig. 43.1) contain a silica resin that binds DNA in the presence of chaotropic salts. The lysate is prepared

G. Barlocchi • F.F. Villa • U. Mastromatteo (✉)
STMicroelectronics, Milan, Italy
e-mail: ubaldo.mastromatteo@st.com

A. D'Amico et al. (eds.), *Sensors and Microsystems: AISEM 2011 Proceedings*, Lecture Notes in Electrical Engineering 109, DOI 10.1007/978-1-4614-0935-9_43,

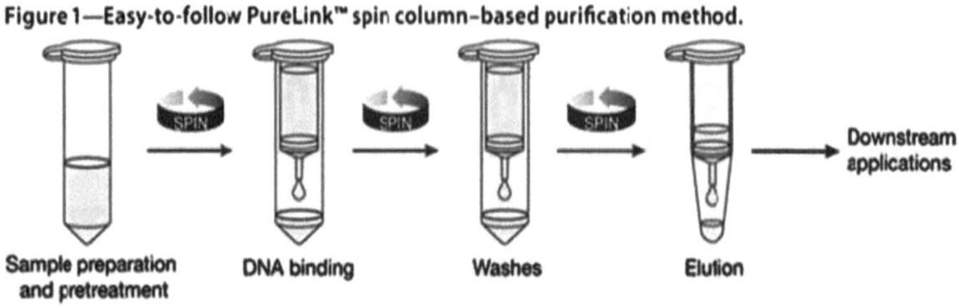

Fig. 43.1 Standard blood DNA extraction procedure

from E. coli cells, yeast cells, mouse tails and mammalian cells and tissues. The cells or tissues are digested with Proteinase K. A detergent is added during lysis to aid in denaturation of proteins and in solubilizing membrane fragments. Any residual RNA is removed by digestion with RNase prior to binding samples to the spin column. The lysate is mixed with Binding Buffer and ethanol to adjust conditions for subsequent DNA binding to the silica-based membrane Spin Column. The DNA binds to the silica-based membrane in the cartridge and impurities are removed by thorough washing with Wash Buffers. The genomic DNA is than eluted in low salt Elution Buffer or water. The method here described is rapid, efficient and thank to Proteinase K, the lysis of cells is obtained without mechanical operations; in spite of that there is some drawback in this procedure because it is manual and time consuming.

2 Automatic Microfluidic System for Real Time PCR Sample Preparation

In order to overcome some of the mentioned limitation, STMicroelectronics is working on a completely integrated system (Fig. 43.2) able to reduce to a minimum the number of steps for the sample preparation, with a suitable fluidic device to be coupled directly on a fast operating Lab On Chip for Real Time PCR.

The system is made of two parts **P1** and **P2** joint together by means of two clips; **P1** is made by plastic material and **P2** is the "disposable cartridge" which hold up a silicon chip with embedded reaction chambers (silicon pockets), heaters and temperature sensors.

2.1 DNA Extraction Sequence

Genomic DNA extraction and purification method is based on the selective binding of DNA to silica-based membrane in the presence of chaotropic salts. Using a siringe pump the whole blood is put inside a special vessel A containing binding buffer and Proteinase K. Proteinase K is used to digest proteins, remove contamination and to inactivate "nuclease", an enzyme, that might otherwise degrade the DNA during

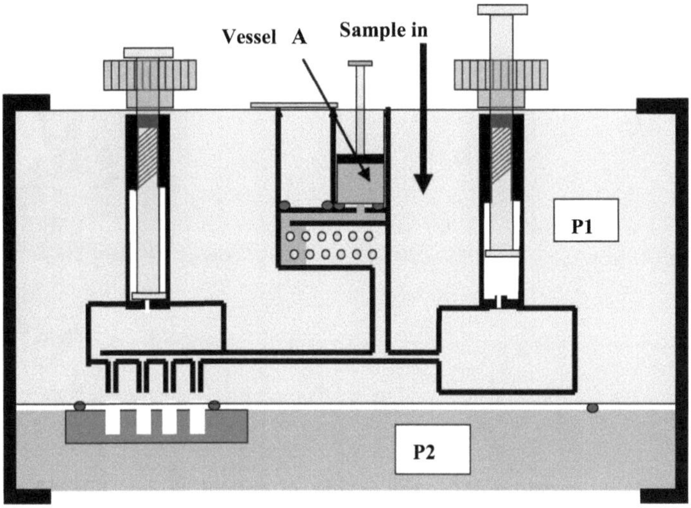

Fig. 43.2 Automatic device system for sample preparation and loading

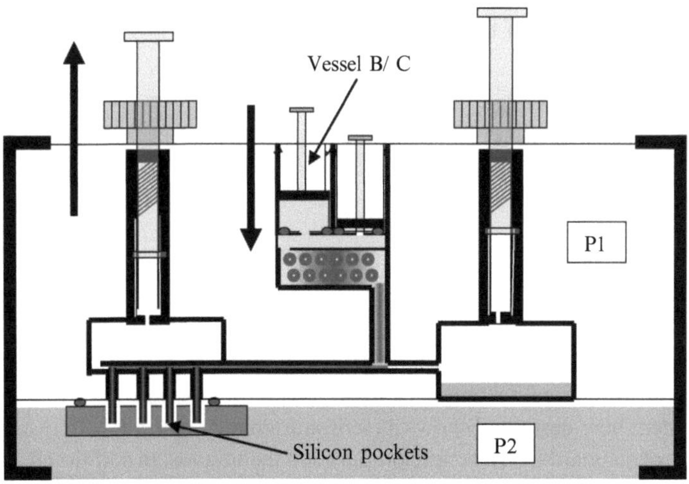

Fig. 43.3 Washing (*Vessel B*) and DNA eluition (*Vessel C*); at the end of the steps DNA eluate will be present inside silicon pockets

purification. A detergent (SDS: Sodium Dodecil Sulfate) is added to improve proteins denaturation and to make soluble the cells membrane fragments. Thereafter by means of the piston of special vessel A the lysate is pushed inside silica-based Spin Column (Fig. 43.2). The DNA binds to the silica-based membrane while impurities are removed (Fig. 43.3) into the waste chamber (bottom right in the Fig. 43.3). Then, the genomic DNA is eluted in low salt Elution Buffer (solution C) and addressed into the silicon chip pockets (bottom left in the Fig. 43.3). At the end of the steps DNA eluate will be present inside silicon pockets of the disposable cartridge P2.

Fig. 43.4 Cartrige (P2) inside an optical reader

The disposable cartridge P2 is put inside an optical reader (Fig. 43.4) and the RT-PCR can be done. A small size prototype of the optical reader instrument has been fully realized in STMicroelectronics (Mechanical, Electronic and Optical components).

Some characteristics of this instrument are:

- Weight: 230 g
- Power supply: 12 V, 0.8A
- Thermal control: ±0.1°C

3 Conclusions

The traditional techniques for DNA sample preparation are manual and require time consuming procedures and need for skilled personnel, so that these techniques are suitable only for clinical testing conducted in central laboratory.

The system here described shows a (semi)automatic device able to overcome the drawbacks of standard DNA sample preparation techniques; in addition it allows the automatic loading of eluate DNA inside silicon pockets. Silicon chip is put on plastic disposable cartridge (P2) and is ready for optical reading during RT-PCR phase.

References

1. Svanvik N, Stahlberg A, Sehlstedt U, Sjoback R, Kubista M (2000) Detection of PCR product in real time using light-up probes. Anal Biochem 287:179–182
2. www.genscript.com; Blood ready TM multiplex PCR system technical manual No. 0174. version 0325208

Chapter 44
Towards MEMS Fabrication by Silicon Electrochemical Micromachining Technology

M. Bassu, L.M. Strambini, and G. Barillaro

In this work, fabrication of Micro Electromechanical Systems (MEMSs) by silicon Electrochemical Micromachining (ECM) technology is demonstrated. High aspect-ratio MEMS structures, with different shape and dimension, consisting of inertial free-standing masses equipped with comb-fingers and suspended by springs from the substrate were fabricated by exploiting advanced features of the ECM technology. ECM fabrication of MEMS structures is here investigated and discussed.

1 Introduction

The field of Micro Electromechanical Systems (MEMSs) has emerged as a technology with significant impact on every day life. MEMSs provide inexpensive means to sense and, in a limited way, control physical, chemical and biological interactions with nature. Today MEMS technology is still far from producing complex structures using low-cost processes with high flexibility. Suspended, deformable micro-mechanical structures have been to date fabricated by mainly using two different methods: anisotropic wet etching tools, such as KOH and TMAH; dry etching tools, such as Deep Reactive Ion Etching (DRIE). Wet anisotropic etching entails the use of a simple set-up at low cost. However, dry etching tools are usually preferred to standard wet techniques, because they offer an enhanced flexibility in obtaining high aspect-ratio structures, which enables the fabrication of a wider range of mechanical elements. On the other hand, such tools are somewhat expensive, and anyway more expensive than wet methods.

M. Bassu • L.M. Strambini • G. Barillaro (✉)
Dipartimento di Ingegneria dell'informazione: Elettronica, Informatica, Telecomunicazioni, Pisa, Italy
e-mail: g.barillaro@iet.unipi.it

A. D'Amico et al. (eds.), *Sensors and Microsystems: AISEM 2011 Proceedings*, Lecture Notes in Electrical Engineering 109, DOI 10.1007/978-1-4614-0935-9_44,

Electrochemical micromachining of silicon in HF-based electrolytes (ECM) is a recently proposed bulk micromachining technology combining advantages of both dry (high-flexibility) and wet (low-cost) traditional micromachining tools [1–3]. Among the main features of ECM the following are worthy to be mentioned: (1) possibility of changing the etching anisotropy (from one to zero); (2) high aspect-ratio of the viable structures. Such features allow an enhanced flexibility in silicon microfabrication to be achieved, even with respect to dry etching technology, and enable high density MEMS fabrication by one single-etching-step. In this work the ECM technology has been pushed to investigate the fabrication of MEMS structures.

2 MEMS Fabrication by ECM

The ECM fabrication of MEMS structures was performed according to the technological steps sketched in Fig. 44.1. An n-type, (100) oriented silicon substrate with a silicon dioxide layer on top was pre-patterned with the layout of the MEMS structure to be fabricated. Standard lithography and subsequent alkaline etching was used to groove a replica of the MEMS layout on top of the silicon surface (seeding points) (Fig. 44.1a and 44.1b). Such a replica has the main purpose of enabling silicon dissolution only in correspondence of the seeding points during the next electrochemical etching (ECE) step.

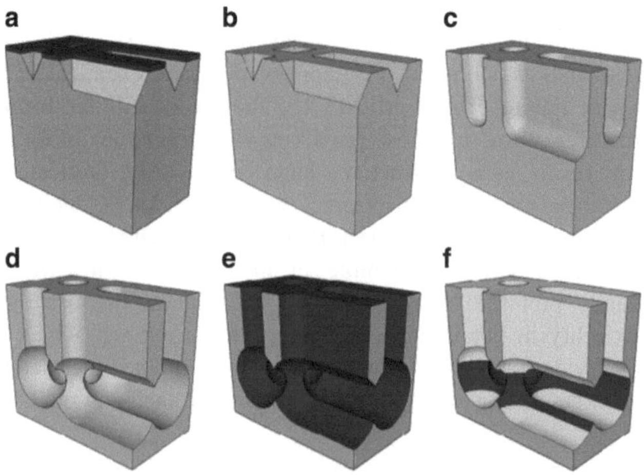

Fig. 44.1 Process flow for the fabrication of free-standing MEMS structures by means of the ECM technology. Definition of an array of holes and trenches (pattern) on the silicon surface (**a**) and (**b**); anisotropic electrochemical etching of the pattern into the substrate (**c**); release of the structures by isotropic electrochemical etching (**d**); dry thermal oxidation (**e**); metallization by sputtering (**f**)

The ECE was performed using a HF-aqueous solution under anodic biasing and back-side illumination of the silicon substrate. The ECE consisted of two distinct phases: (1) an initial anisotropic phase, and (2) a final isotropic phase.

The anisotropic phase was exploited to deep etch the layout (combination of pores and trenches) into the silicon substrate and ended as soon as a given etching depth was reached (Fig. 44.1c). At this stage the etched structures are still anchored to the substrate both in the horizontal direction (through actuation springs) and in the vertical direction (through the bottom of the structure itself). The next isotropic phase was exploited to etch the fabricated structures at their bottom and, in turn, release them from the substrate (Fig. 44.1d). At this stage the etched structures are free-standing and only anchored to the substrate through actuation springs. The etching anisotropy degree was tuned by properly setting the etching current density during the electrochemical etching step. Once free-standing structures were produced, thermal growth of silicon dioxide (100–200 nm thick) (Fig. 44.1e) followed by conforming sputtering deposition of metal (100–200 nm thick) (Fig. 44.1f) were performed for the formation of insulated, metal electrodes for electrical actuation and/or capacitive sensing, according to the well-known SCREAM process [4].

3 Experimental Results and Discussion

In this work, the electrochemical etching step is thoroughly discussed and experimental results on silicon microstructure fabrication are given.

Fabrication of high aspect-ratio structures requires an accurate control of the etching along the vertical (perpendicular to silicon surface) direction as the etching progresses. In the ECM technology this is obtained by finely tuning both the etching current density J_{etch} and the anodization voltage V_{etch} during the anisotropic phase of the etching. A characteristic peak J_{ps} for a given voltage V_{ps} in the current density-voltage curve of the Si-HF electrochemical system allows to identify the ECM working region: $J_{etch} < J_{ps}$ and $V_{etch} > V_{ps}$. The J_{etch} value is chosen as $J_{ps}P$, being P the porosity value (etched silicon volume with respect to total silicon volume under etching) to be obtained for the structure under fabrication. The value of J_{etch} is changed as the etching progresses so as to compensate for the change of the critical current density J_{ps} due to the diffusion-limited transport of HF inside the narrow etched grooves, and to maintain a constant porosity value over the whole etching depth. The V_{etch} value is also changed as the etching progresses in order to obtain a better control on the shape of the etched microstructure. The release of the structures etched during the anisotropic phase, which are still anchored to the silicon substrate at their bottom, is obtained by properly increasing the etching current density in order to make the etching becoming isotropic, thus allowing under-etching of silicon to be fully accomplished at the bottom of the structures.

Figure 44.2 shows two examples of MEMS structures consisting of free-standing inertial masses, one of which equipped with comb-finger batteries, connected to an anchoring structure by serpentine springs. The inertial masses consist of an array of

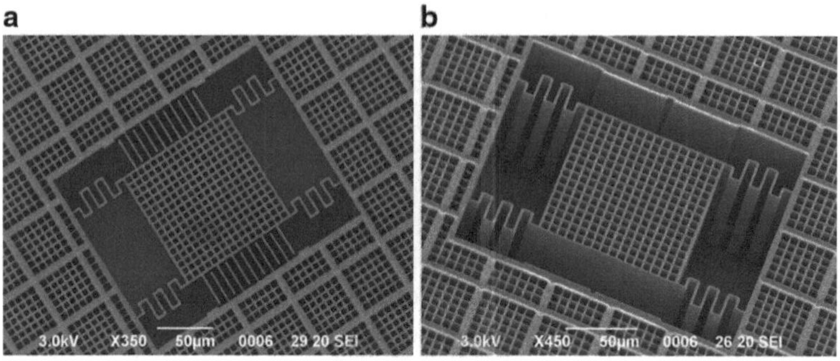

Fig. 44.2 SEM images of mass-spring structures fabricated by electrochemical micromatching technology. Top view of a structure equipped with serpentine springs and comb-fingers (**a**) and tilted view of a similar structure without comb-fingers (**b**), both featuring a depth of 98 μm

holes with side of 1.5 μm and pitch of 6.8 μm. Comb-fingers as well as serpentine and beam springs have thickness of 2 μm and height up to 98 μm. The anchoring structure consists of a two-dimensional array obtained by repetition of a basic block, the latter made up of a 6 x 6 square array of holes with same geometrical features as inertial masses.

4 Conclusions

In this work, fabrication of MEMS structures by ECM silicon technology has been experimentally proven to be feasible. Free-standing inertial masses equipped with comb-finger batteries and suspended from the substrate by high aspect-ratio springs were for the first time successfully fabricated by ECM technology. Experimental results establish the ECM as a low-cost wet-etching technology with flexibility comparable with that of high-expensive dry-etching tools.

References

1. Barillaro G, Bruschi P, Diligenti A, Nannini A (2005) Fabrication of regular silicon microstructures by photoelectrochemical etching of silicon. Phys Stat Sol (c) 2:3198–3202
2. Barillaro G, Nannini A, Piotto M (2002) Electrochemical etching in HF solution for silicon micromachining. Sensor Actuat A 102:195–201
3. Barillaro G, Diligenti A, Benedetti M, Merlo S (2006) Silicon micromachined periodic structures for optical applications at $\lambda = 1.55$ μm. Appl Phys Lett 89:151110
4. Zhang ZL, MacDonald NC (1992) A RIE process for submicron, silicon electromechanical structures. J Micromech Microeng 2:31–38
5. Lai SL, Johnson D, Westerman R (2006) Aspect ratio dependent etching lag reduction in deep silicon etch processes. J Vac Sci Technol A 24:1283–1288

Chapter 45
Development of a SOLT Calibration Setup for SAW Sensor Characterization

N. Donato and D. Aloisio

In this paper is reported the development and validation of a calibration kit, made of standard loads, specifically dedicated to SAW resonators, with TO39 package. By using this calibration kit it is possible to perform a SOLT (Short Open Load Thru) VNA (Vectorial Network Analyzer) calibration, allowing the deembedding of the package parasitic effects in a frequency range up to 500 MHz. The developed calibration kit was employed in the characterization of a SAW sensor based on Poly (methacrylic acid) (PMA) as sensing material for ethanol monitoring.

1 Introduction

Over the last decade, Surface Acoustic Wave (SAW) sensors have been of great interest in chemical applications. SAW sensors permit an highly sensitive detection of gas molecules because the acoustic energy is highly confined into the near-surface region of a piezoelectric substrate, causing extremely sensitive characteristics of acoustic wave propagation to any surface perturbation [1]. The simplest way to develop a SAW sensor is to employ a commercial SAW resonator as a transduction substrate, and to deposit on it an appropriate sensing film. The operating frequencies span from a few dozen to a few 100 MHz. On the other hand, it is well-known that the mass sensitivity of such devices is directly proportional to the square of its operating frequency, and SAW sensors for organic vapours, for instance, are usually mass sensitive [2].

An improvement in testing the properties of SAW sensors can be accomplished by reducing all parasitic effects due to the package of the device [3].

N. Donato (✉) • D. Aloisio
Department of Matter Physics and Electronic Engineering, University of Messina, Messina, Italy
e-mail: ndonato@unime.it

A. D'Amico et al. (eds.), *Sensors and Microsystems: AISEM 2011 Proceedings*, Lecture Notes in Electrical Engineering 109, DOI 10.1007/978-1-4614-0935-9_45,

2 Experiments

In this work is reported the development and validation of a calibration kit, made of standard loads, specifically dedicated to SAW resonators, with TO39 package. By using this calibration kit it is possible to perform a SOLT (Short Open Load Thru) VNA (Vectorial Network Analyzer) calibration, allowing the deembedding of the package parasitic effects.

This calibration is performed by inserting three standards: a Short, an Open, a Load (50 Ω), for each side (left and right) of the characterization setup and a Thru standard to connect both sides together. The calibration kit was developed for each side (left and right) of the measurement system.

The standards were developed by means of RF surface mounting devices in "0805" format, (in Fig. 45.1 is reported the schematic and the photograph of the standards developed for the right side). In Fig. 45.2 are reported the S_{11} and S_{21} parameters for standards load before and after calibration, i.e. by considering as reference plane at first the SMA connectors of the developed test fixtures (plane B), then considering as reference plane the intrinsic device. It can be seen the phase rotation due to the different electrical length. In this way it is possible to analyse the "intrinsic" behaviour of crystal and IDT during the investigation on SAW devices as sensors. By removing the parasitic effects it is possible to achieve a greater accuracy in the electrical characterization (by measuring S parameters), and to develop simpler circuital networks allowing the electrical modeling of the device near the resonance frequency. The SAW sensor with TO39 package is placed in a test fixture, made in our laboratory with a 25 N-30 high frequency substrate and provided with two 50 Ω SMA female connectors. The realized test fixture and the calibration kit were validated by measuring commercial SAW resonators in a

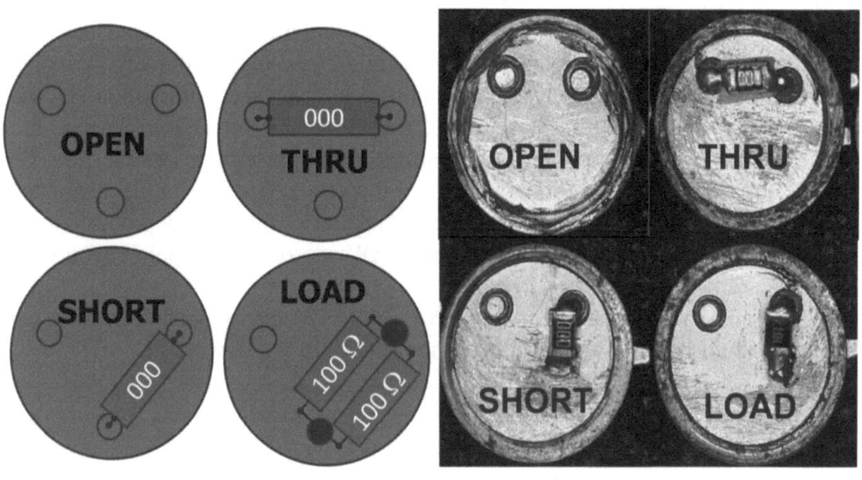

Fig. 45.1 Schematic and photograph of the developed Cal-Kit (*right side*)

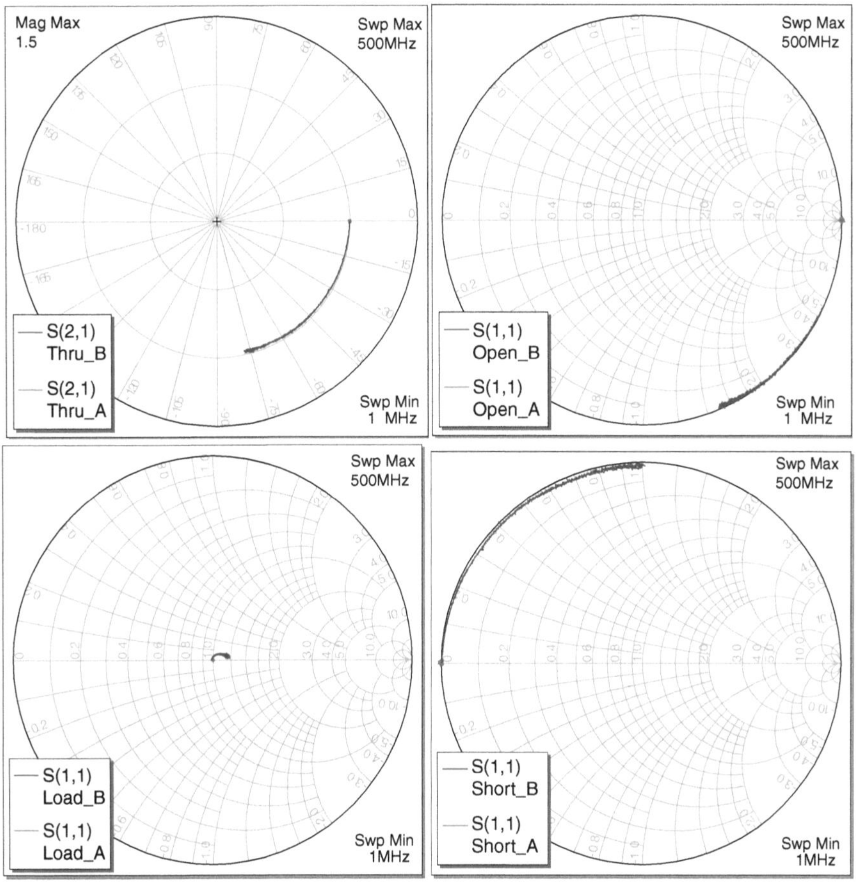

Fig. 45.2 S11 and S21 parameters of standards before (_B) and after (_A) calibration

frequency range from 1 to 500 MHz, then they were employed in the electrical characterization of a 423.3 MHz sensor.

3 Results

The calibration system was validated by measuring the S parameters of SAW resonators working at 235.2, 418, 423.3, 433.92 MHz respectively (not reported) with Agilent 8753ES VNA and with a low cost VNA based on Analog Devices AD9859 (direct digital synthesizer, DDS). The measurement procedure was then employed in the characterization of a SAW sensor working at 423.3 MHz by measuring the S parameters and then, the resonance frequency shift.

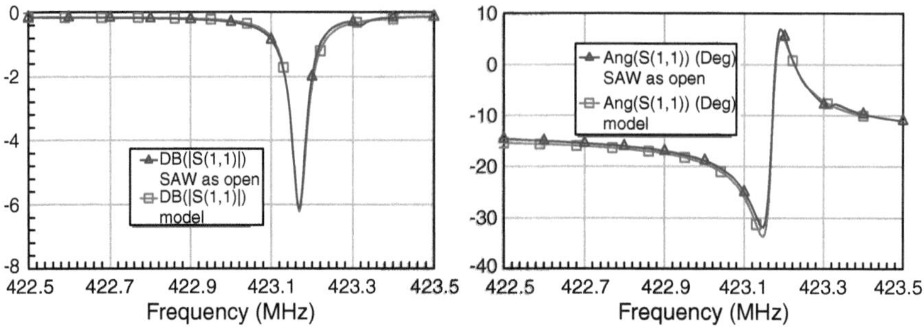

Fig. 45.3 Comparison between measured data and modeled ones

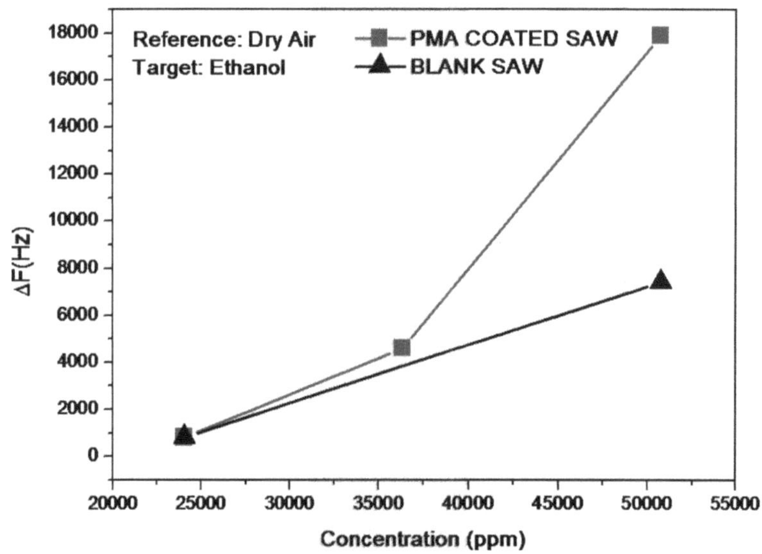

Fig. 45.4 Resonance frequency shift vs ethanol concentration

In Fig. 45.3 is reported the comparison between the measurement data and the modeled ones of the SAW resonator working at 423.3 MHz, as opened, before the sensing film deposition. The model was extracted and computed my means of commercial electrical CAD tools. It can be seen the good agreement between the model and measurements. After the characterization of the "blank", a sensing film was deposited on the device, by drop coating a solution of PMA in water.

Once developed, the sensor was inserted in a test chamber, mounted on a automated characterization system able to monitor and control the flow of the sample gas mixtures. It can be seen that by increasing the ethanol concentration the resonance frequency decreases (Fig. 45.4).

4 Conclusions

In this paper it is reported the development of a calibration kit for SAW devices on TO39 packages. The SOLT calibration was validated by measuring commercial devices and an home made ethanol sensor. The sensing device was developed by depositing a PMA layer on a commercial SAW resonator working at an resonance frequency of 423.3 MHz. These preliminary measurements show the sensing properties of the PMA towards ethanol, allowing the development of SAW sensors.

References

1. Penza M, Cassano G, Aversa P, Antolini F, Cusano A, Cutolo A, Giordano M, Nicolais L (2006) Recognition of organic solvents molecules by simultaneous detection using SAW oscillator sensors and optical fiber devices coated by Langmuir-Blodgett cadmium arachidate films. IEEE Trans. on Ultrasonics, Ferroelectrics, and Frequency Control, 53(8):1493–1502
2. Penza M, Aversa P, Cassano G, Wlodarski W, Kalantar-Zadeh K (2007) Layered SAW gas sensor with single-walled carbon nanotube-based nanocomposite coating, 127(1):168–178
3. Fischerauer G, Gogl D, Weigel R, Russer P (1994) Rigorous modeling of parasitic effects in complex SAW RF filters, "Microwave Symposium Digest, (1994), IEEE MTT-S International, 2:1209–1212, doi: 10.1109/MWSYM.1994.335563

Chapter 46
A Very Large Dynamic Range Integrated Interface Circuit for Heterogeneous Resistive Gas Sensors Matrix Read-Out

Fabrizio Conso, Marco Grassi, Piero Malcovati, and Andrea Baschirotto

In this chapter we present a low-cost multiplexed interface circuit based on a resistance-to-time converter suitable for the complete read-out of a 5×5 gas sensors matrix. The front-end circuit includes a bias branch, a continuous time integrator, a couple of discriminators, a control logic and 20 sets of analog multiplexers. Low-cost feature is achieved by means of the fact that the interface circuit is shared between different read-out configurations and that no calibration or bulky auto-ranging are mandatory. The worst case linearity error is about 0.3%, when using a single fixed integrating capacitor. The presented read-out circuit performance satisfies the requirements of most environmental gas monitoring applications for which a precision better than 1% is enough in order to be sure to detect every gas type of interest with adequate accuracy.

1 Introduction

Today's air pollution, with particular focus on big towns, and people health and safety related to the exposure of dangerous gases are among main concerns of all national governments and institutions over the world. For this reason worldwide and local legislative guidelines have been issued and continuously updated to rise the quality of life thus reducing health risks*. In order to apply defined rules for the quality of air in towns and to limit the exposure to dangerous gases in controlled and

*Work carried out within project PRIN *20085AJSEB* funded by MIUR, Italy

F. Conso • M. Grassi (✉) • P. Malcovati
Department of Electrical Engineering, University of Pavia, Pavia, Italy
e-mail: marco.grassi@unipv.it

A. Baschirotto
Department of Physics, University of Milano Bicocca, Milano, Italy

A. D'Amico et al. (eds.), *Sensors and Microsystems: AISEM 2011 Proceedings*, Lecture Notes in Electrical Engineering 109, DOI 10.1007/978-1-4614-0935-9_46,

free environments, several commercial dedicated systems are available, which are mainly based on high-cost laboratory instruments modules. They are usually installed in suitable fixed places or carried around by means of dedicated vehicles. While legislative guidelines are getting more and more severe, the number of required systems for air monitoring and relative precision are growing everyday and the associated expense is not negligible anymore compared to other duties of local administrations and security agencies. Newer required devices, either fixed or portable, demand, as mentioned, for higher accuracy and reliability as well as lower maintenance costs. Lower cost and higher portability of such devices may be pursued thanks to the exploitation of a combination of micromachining technology for the gas-sensing element and traditional integrated CMOS technology for the read-out circuit, also called "micromodule approach". Indeed, micromachined sensors may be fabricated with embedded heater and thermometer, reaching the operative temperature in few tens of milliseconds with a power consumption of the order of milliwatts, while the read-out integrated circuit (IC), operating at ambient temperature, may be fabricated without the restrictions imposed by the sensor related technology, thus allowing the maximum overall performance.

In these IC-based instruments the digitized outputs of several different unitary elements arranged in a sensor matrix are properly processed with dedicated digital pattern recognition algorithm software. This ensures that even if each one of the N sensing devices is optimized to detect a particular kind of gas, all of them provide important information also on the concentration of other gas types. In particular, these microfabricated sensors are realized by thick film deposition of metal oxide sensitive materials on standard dielectric layers, such as silicon dioxide or silicon nitride, and their operating principle is related with chemi-adsorption and charge transfer processes between gas molecules and MOX film, which causes a simple electrical resistance variation of the gas sensing element.

The grid [1] to read-out, depicted in Fig. 46.1, is composed of 5×5 micro-membranes, each equipped with four electrodes to contact the sensitive material deposited on top of the membrane itself, a heater and a temperature sensor. The four electrodes of each micromembrane are directly connected to the electrodes of the four adjacent membranes, thus creating the actual grid. To further reduce the number of interconnections between the grid and the periphery all heaters and temperature sensors of each row are connected in series and, hence, the membranes of each row operate at the same temperature. In the case of this work, a strong cost reduction is guaranteed if a single interface circuit is able to process the analog information from all the rows and columns of the matrix. This requires a multiplexing technique.

Moreover, wide dynamic range performance is also needed because the sensor resistance value (R_{sens}) may vary through several decades, being the combination of three independently variable components: the baseline R_{bl}, which mainly depends on the fabrication technique, the deviation from the baseline ΔR_{bl}, due to technological and aging spread and temperature, as well as the resistance variation ΔR_{gas} which depends on gas concentration, negative for most gas types. This variation range can be as large as a couple of decades. At the present state-of-the-art in SnO_2 thin oxide sensor manufacturing, the baseline usually varies from about 1 MΩ to 1 GΩ and the

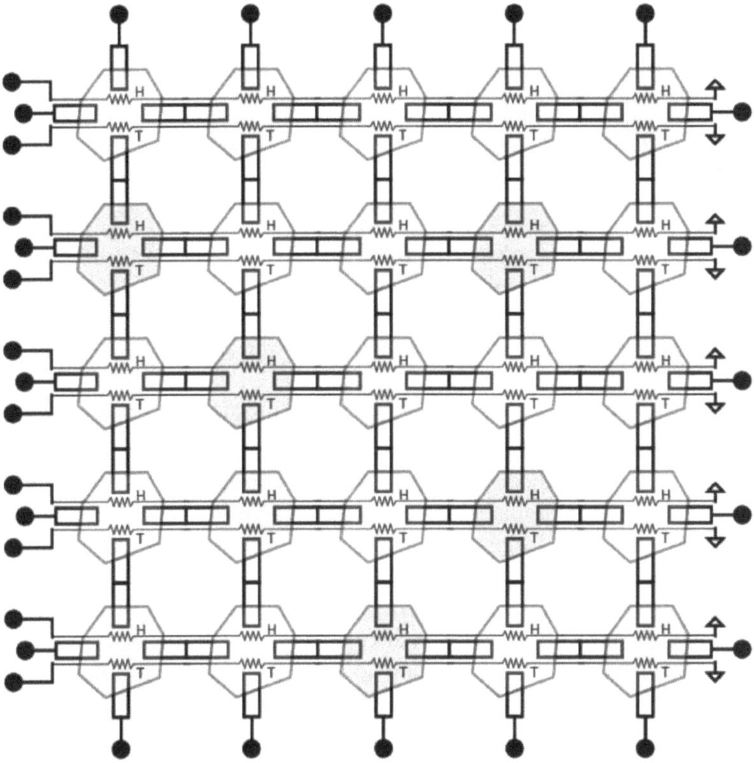

Fig. 46.1 Schematic diagram of the 5 × 5 gas sensor grid to be read-out

sensor resistance has to be measured with a precision better than 1%, in order to be sure to detect every gas type with enough accuracy for most environmental gas monitoring applications. Thus the overall dynamic range required for the interface circuit connected to a singular element may be as large as more than five decades, according to $R_{sens} = R_{bl} + \Delta R_{bl} + \Delta R_{gas}$, e.g. ≈ [10 kΩ–1 GΩ].

2 Sensor Resistance Read-Out: State of the Art

Assumed that the value to be read-out is a resistance, suitable circuit topologies could be the resistance to voltage and the resistance to time converter. Furthermore, in order to take into account such a large dynamic range, different approaches can be used. In literature many techniques are described in detail, such as the traditional [2] resistance controlled oscillator approach, which allows the measurement of a large resistance range, but with low linearity because the sensor parasitic capacitance is exposed to a large voltage swing. Another solution exploits logarithmic compression of the voltage signal obtained as difference of the voltage drops over

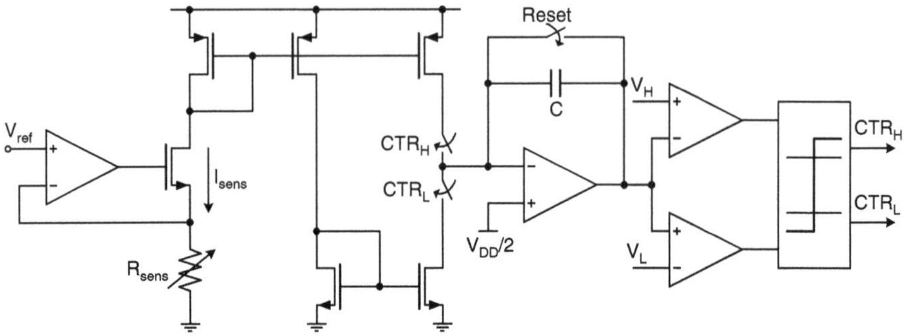

Fig. 46.2 Schematic of the enhanced resistance-to-time converter reported in literature

two diodes, one biased with the sensor current and the other with a precise reference one [3]. However, this method does not achieve very high linearity over several decades due to intrinsic diode characteristic even considering compression. A last architecture is the programmable trans-resistance amplifier, which is still inspired to high precision laboratory instruments. As mentioned in the introduction, this last kind of approach is quite expensive: in fact it requires an auto-ranging circuit for scale selection, as well as a DSP to perform coarse and fine calibration in order to cancel any inter-range offset and gain errors [4].

So far, the most promising solution to achieve such a large dynamic range without the need of calibration has been proven to be the enhanced oscillator approach, which may also be found in literature [5]. In this architecture, as shown in Fig. 46.2, the voltage across the sensor is kept constant, and equal to the reference voltage V_{ref}, thus granting a correct sensing operation. The current obtained is then injected into or out from the input of a single-ended integrator by means of precision current mirrors. The output of the oscillator is squared by a set-reset flip-flop and used to drive the current mirror itself, while two comparators bound the integrator output signal in a given range, obtaining in this way a triangular wave with a period proportional to R_{sens}.

This solution achieves a very good resistance to time conversion, also in terms of linearity performance. The main issue in this case is that it would not be possible to interconnect the proposed circuit to more than a single sensor without paying in terms of linearity drop due to unavoidable parasitic resistance (both R_{on} and R_{off}) of the multiplexer related analog switches that is necessary to select which sensor resistance has to be converted every time.

3 Developed Multiplexed Enhanced Resistance-to-Time Converter

The developed resistance to time converter circuit is an optimized version of the enhanced oscillator circuit, which has been designed in a 0.35 μm CMOS technology with a nominal power supply of 3.3 V. This new architecture exhibits a lower

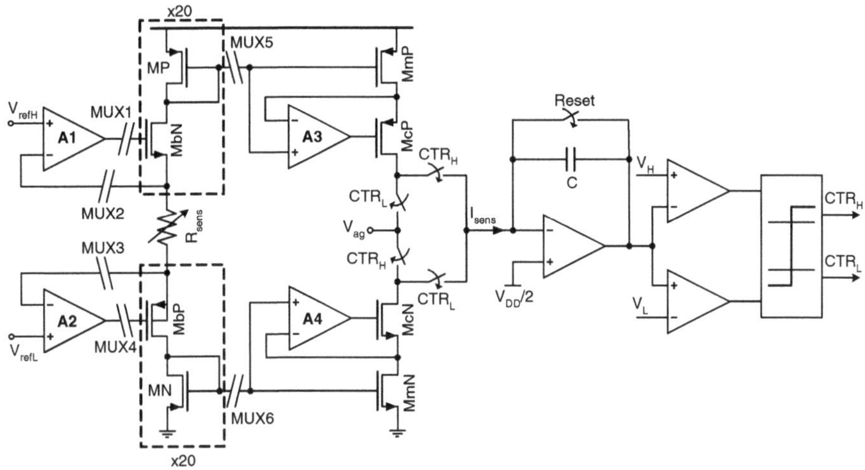

Fig. 46.3 Schematic of the developed resistance to time converter with high linearity multiplexer

power consumption, while improving the allowed dynamic range. Moreover, it allows the cancellation of the parasitic effects of the additional multiplexer required to drive the sensor matrix. Figure 46.3 shows the schematic of the proposed circuit. The resistor R_{sens} is biased with a constant buffered voltage, which is set to 0.5 V. Source follower connected transistors MbN and MbP have been designed with a large W/L ratio to keep their overdrive very low, thus preventing saturation of the output stages of amplifiers A1 and A2 when I_{sens} is maximum. Transistors MbN and MbP still have about 300 mV of V_{DS} saturation margin even in worst operation conditions, granting the development of this architecture. It is worth to note that, due to the unavailability of n-MOS transistors with insulated substrate in the employed technology, the reference voltages can not be symmetrical with respect to VDD/2, keeping into account of the larger threshold voltage of MbN with respect to its p-channel counterpart. On the other hand, transistors McN and McP have been designed to have a small W/L ratio, to ensure that, with the lowest I_{sens} and maximum temperature, the output voltage of amplifiers A3 and A4 is sufficiently away from the power supply and ground respectively. In order to cancel the parasitic effects of the multiplexers driving the 5×5 sensor matrix, they have been connected as shown in Fig. 46.3, replicating the bias branch 20 times, one for each peripheral output pin of the matrix. We remind that overall sensor response carries additional information also depending on the current direction and for this reason peripheral pins must have the possibility to be swapped in read-out operation. The current of the bias branch I_{sens} is injected directly into or out from the input of a single ended integrator by means of a 1:1 current mirror. In this way we obtain an integrating current in the range 0:5 nA–0:5 mA, which is a good compromise between power consumption, linearity and architecture complexity. Since the integrator output is bounded to the range $\Delta V = 1.15$ V–2.15 V, the resulting triangular wave will have, according to a period that varies in the range 400 ns–400 ms, an oscillator frequency in the range 2.5 MHz–2.5 Hz, for an integrator capacitance of 100 pF. This value has been chosen as a trade-off between conversion speed and power consumption of the integrator amplifier. As explained in literature, the ratio between a reference counter, whose clock frequency is set at the midrange value $f_{mid} = (f_{min} \cdot f_{max})^{1/2}$ and the resistance dependent one represents the numeric converted value of R_{sens}. The measurement ends when the slower of the two counters reaches a given value N^*, which is enough to achieve the desired accuracy, i.e. $N^* = 256$ for 8-bit local resolution. In typical operating mode, f_{mid} is set to 2.5 kHz.

Considering a maximum singular conversion time of $256T_{max}$, 102.4 s is the complete conversion time of the whole sensor matrix required to execute $19 \times 20 = 380$ measurements. An overall conversion time per array scan of about 100 s may not be acceptable. For this reason an optional functionality has been introduced to reduce by a factor down to 100 times the conversion of the highest resistance values, while introducing a small almost constant delay in the conversion time of all measurements. With this method, the circuit executes a first rough conversion which lasts at maximum 12.8 ms (32 periods of the reference clock) and, if saturation occurs in the direction of high resistance values, the integrating capacitance is lowered to 1 pF, instead of initial value of 100 pF, boosting the oscillator frequency. Unfortunately, this capacitance value cannot be reduced further to avoid the introduction of a significant non linearity due to mismatches. Indeed, for the used technology, the poly-poly capacitor matching is given by: $\Delta C/C = 0.47 \cdot 10^{-6}/(W \cdot L)^{1/2}$, which results in a 0.4% matching for a minimum capacitance of 1 pF.

4 Simulation Results

The 5×5 gas sensor matrix read-out circuit has been developed in 0.35 μm CMOS technology. Figure 46.4 shows the relative linearity error in sensor resistance estimation at transistor level simulations with the integrator capacitance equal to 100 pF. The circuit has been simulated with a power supply voltage of 3.3 V and a temperature of 80°C, which represent the worst case condition.

The circuit exhibits a linearity precision better than 0.1% for resistance values up to 700 MΩ. For higher resistance values the leakage currents, especially from the reset transistor and from the chip pads ESD protections, whose effect is no longer negligible, induce a proportional increase in the linearity error. Finally, with the additional nonlinear term introduced by the capacitor mismatch the overall interface non-linearity becomes 0.5%, with a maximum conversion time of 0.5 s, or remains 0.35% for a maximum conversion time of 2 s. The total current consumption of the circuit, in typical condition and for low sensor resistance values is approximately a linear function of the sensor conductance itself and varies in the range 1.3 mA – 2.3 mA.

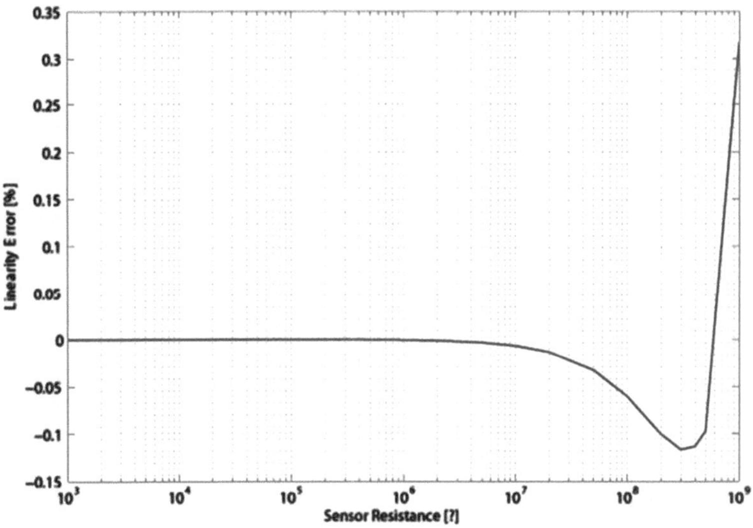

Fig. 46.4 Simulated linearity relative error in sensor resistance value read-out

5 Conclusions

A 8-bit accuracy, six-decade range resistive gas-sensors matrix read-out circuit has been developed and simulation results have been reported. Thanks to the employment of a resistance-to-time conversion with the oscillator decoupled from the sensing device and suitable analog low insertion loss (either in terms of R_{on} and R_{off}) multiplexers, the circuit in transistor level simulations achieves a worst-case precision of about 0.35% over a range of six decades (1 kΩ–1 GΩ) in normal operation mode, without requiring any calibration nor complicated autoranging procedure.

References

1. Grassi M, Malcovati P, Francioso L, Siciliano P, Baschirotto A (2007) Integrated interface circuit with multiplexed input and digital output for a 5×5 SnO_2 thick film gas sensor matrix. Sensor Actuator B 132:568–575, Elsevier
2. Flammini A, Marioli D, Taroni A (2004) A low-cost interface to high value resistive sensors varying over a wide-range. Trans Instrum Meas 53:1052–1056, August 2004, IEEE
3. Barrettino D, Graf M, Song WH, Kirstein K, Hierlemann A (2004) Hotplate based monolithic CMOS microsystems for gas detection and material characterization for operating temperatures up to 500°C. J Solid State Circuits 39:1202–1207, IEEE
4. Grassi M, Malcovati P, Baschirotto A (2007) A 160-dB equivalent dynamic range auto-scaling interface for resistive gas sensors arrays. J Solid-State Circuits 42(200):518–528, IEEE
5. Grassi M, Malcovati P, Baschirotto A (2007) A 141-dB dynamic range CMOS gas-sensor interface circuit without calibration with 16-bit digital output word. J Solid-State Circuits 42:1543–1554, IEEE

Chapter 47
Design of an Electronic Oscillator Based on an On-Chip MEMS Resonator Aimed at Sensing Applications

F. Pieri, V. Russino, and P. Bruschi

We present the design of an integrated electronic oscillator aimed at the detection of the response of an integrated MEMS resonant mass sensor. The resonator and oscillator are designed to be fabricated on a CMOS-compatible bulk micromachining technology which allows the coexistence of MEMS components and integrated circuits on the same chip. The resonator is targeted at the development of low-cost integrated smart biosensors for the detection of diagnostic markers in a POCT (point-of-care testing) context. The oscillator is based on a standard positive feedback topology, using the MEMS resonator as a two-port, frequency selective element. The main design issues are the low input impedance of the resonator and its significant attenuation. ELDO^{TM} simulations of the designed circuit, based on an equivalent model of the resonator extracted from experimental data, were performed, and the results are presented and commented.

1 Introduction

Resonant mass sensors are widely used in sensing and biosensing applications, owing to their high mass-to-frequency sensitivity. While most applications in the biosensing field are based on the piezoelectric quartz resonator (Quartz Crystal Microbalance or QCM), resonant MEMS sensors have been proposed as well. The output frequency shift can be measured with different approaches, ranging from measurement of the full impedance spectrum, to the inclusion of the resonator in an electronic oscillator, to more complex schemes based on PLL locking [1].

F. Pieri • V. Russino • P. Bruschi (✉)
Dipartimento di Ingegneria dell'Informazione: Elettronica, Informatica, Telecomunicazioni, Università di Pisa, Pisa, Tuscany, Italy
e-mail: p.bruschi@iet.unipi.it

A. D'Amico et al. (eds.), *Sensors and Microsystems: AISEM 2011 Proceedings*, Lecture Notes in Electrical Engineering 109, DOI 10.1007/978-1-4614-0935-9_47,

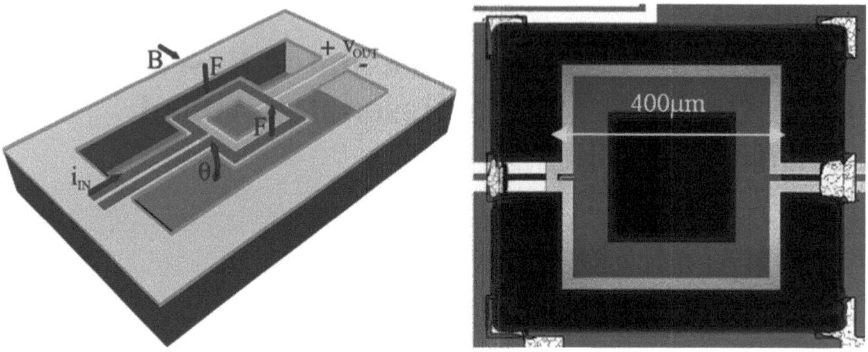

Fig. 47.1 Sketch of the fabricated MEMS resonator (*left*); optical top view (*right*)

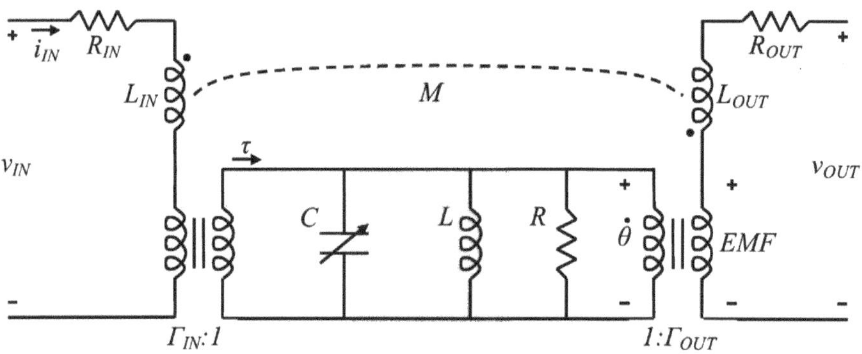

Fig. 47.2 MEMS resonator equivalent circuit

In this work, we present the design of an integrated electronic oscillator circuit based on an on-chip MEMS resonator. The resonator is fabricated in a CMOS-compatible bulk micromachining technology [2]. To allow the implementation of an on-chip oscillator, two general purpose operational amplifiers were also included on the resonator chip. The resonator operation is based on magnetic transduction. Two integrated inductors are embedded in a suspended dielectric membrane (Fig. 47.1), which is driven into torsional motion by a sinusoidal current injected in the first inductor. The second inductor is used to detect the motion thanks to the electromotive force induced at its terminals. The resonator is designed to be used as the transduction component of a mass biosensor based on the microbalance principle.

To this purpose, the surface of the resonator is functionalized to bind specifically with a biomolecule of diagnostic interest. The consequent variation in the membrane mass can be detected as a variation in its mechanical resonance frequency, following the well known microbalance principle. The overall device can be modeled by an equivalent circuit (Fig. 47.2), whose central section models the mechanical part, while the two transformers model the electromechanical transduction. The electrical parameters of the equivalent circuit, which are required for the design and simulation of the oscillator, were extracted from the measurement of the resonator frequency response.

2 Oscillator Circuit

The resonator is used in the feedback loop of the oscillator shown in Fig. 47.3. The cascade of two op-amp based non-inverting amplifiers provides the required loop gain. The op-amps are general purpose, rail-to-rail, class AB amplifiers [3], integrated on the same chip as the resonator. The microbalance is inserted into the feedback path, with the input section connected in series with the group C_2/R_6. It can be easily shown that the voltage of the bias generator V_B is transferred across R_6. By this connection, the output stage of OP2 is forced to source the current V_B/R_6, which has been set to 1 mA. As a result, the class AB output stage is biased in a region where the small signal output resistance is much lower than in the quiescent state. This configuration was necessary to drive the very low resistance of the microbalance input section reducing the attenuation due to the loading effect of the latter. The nominal voltage gains of the first and second stage are 400 and 3, respectively. The necessity of using two op-amps derived from the limitation in the gain-bandwidth product of the available amplifiers (around 2 MHz), combined with the high insertion loss of the microbalance.

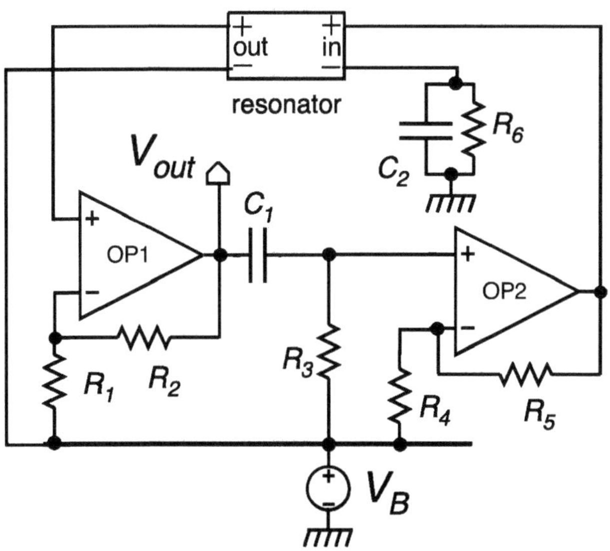

Fig. 47.3 Schematic view of the oscillator circuit

3 Simulation Results

The circuit has been simulated using the ELDO™ (Mentor Graphics) electrical simulator, configured with the device models of the same process used for the microbalance and amplifier fabrication (STMicroelectronics' BCD6s). The microbalance was modeled with the equivalent circuit mentioned above. The feedback loop was preliminarily cut at the OP1 input to estimate the open loop frequency response. A phase lead of nearly 45°, required to compensate for the phase lag introduced by the amplifiers, is provided by capacitor C_2. Transient simulations in closed loop configuration show that a test current pulse (100 nA–100 ns) injected into the OP1 inverting input induces an oscillation at the resonance frequency of the resonator (around 30 kHz), as expected (Fig. 47.4). The steady condition was reached after about 20 ms. The output voltage reaches an amplitude of about 60 mV peak-to-peak. The output voltage at the OP2 output is larger (around 100 mV peak-to-peak) but significantly distorted. The reason is saturation of the OP2 output current due to the very low impedance of the load (12 Ω). Therefore saturation is the self-limiting mechanism for the oscillations. The result of detailed simulations on the circuit will be used to design and develop a practical prototype.

Acknowledgments The authors thank STMicroelectronics for allowing access to the BCD6s technology. This work was partly financed by the Italian Ministry of Education, University and Research under a PRIN grant.

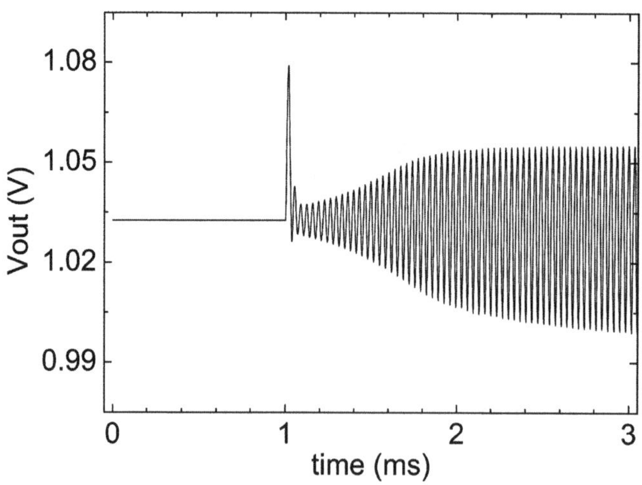

Fig. 47.4 Simulation of the oscillation start-up transient following a 100 nA–100 ns current pulse injected at the 1 ms instant

References

1. Arnau A (2008) A review of interface electronic systems for AT-cut quartz crystal microbalance applications in liquids. Sensors 8:370–411
2. Paci D, Pieri F, Toscano P, Nannini A (2008) A CMOS-compatible, magnetically actuated resonator for mass sensing applications. Sensor Actuator B 29:10–17
3. Hogervorst R, Tero J, Eschauzier R, Huijsing J (1994) A compact power-efficient 3 V CMOS rail-to-rail input/output operational amplifier for VLSI cell libraries. IEEE J Solid-State Circuits 29:1505–1513

Chapter 48
An Analog Automatic Lock-In Amplifier for the Accurate Detection of Very Low Gas Concentrations

Andrea De Marcellis, Giuseppe Ferri, Arnaldo D'Amico, Corrado Di Natale, and Eugenio Martinelli

We propose here a new analog lock-in amplifier to be utilized in sensor interfaces for the detection of very low quantity of dangerous gases. When compared to other commercial systems in the literature, the proposed scheme shows an automatic operation, consisting in the self-alignment of the relative phase between input and reference signals. This functionality is continuously guaranteed, both at power-on and for any variation of the input noisy signal phase and amplitude during the working time. The proposed lock-in has been designed to work at a specified reference frequency (77 Hz), suitable for gas sensor applications and that avoids interferences with 50 Hz net frequency and its harmonics. The system has been tested using the carbon monoxide as gas to be revealed. With respect to the simple resistive gas sensor interface implemented by a resistive voltage divider, the improvement given by the proposed lock-in amplifier for the system sensitivity is of a factor of about 80, while the resolution, starting from about 5 ppm, has been enhanced to a theoretical value of about 0.05 ppm.

1 Introduction

The lock-in technique measures the magnitude of a signal, buried into noise, in a very narrow frequency bandwidth, while rejects all the components of the signal outside it. It shows better performance than a simple filtering operation, because of the automatic tracking that allows lock-in amplifiers to give effective quality

A. De Marcellis (✉) • G. Ferri
Department of Electrical and Information Engineering, University of L'Aquila, L'Aquila, Italy
e-mail: andrea.demarcellis@ing.univaq.it

A. D'Amico • C. Di Natale • E. Martinelli
Department of Electronic Engineering, University of Tor Vergata, Roma, Italy

A. D'Amico et al. (eds.), *Sensors and Microsystems: AISEM 2011 Proceedings*, Lecture Notes in Electrical Engineering 109, DOI 10.1007/978-1-4614-0935-9_48,

factor Q values (a measure of filter selectivity) over 100,000, whereas a normal band-pass filter does not give a Q greater than 50.

In sensor interface design, sometimes it is necessary to maximize the system resolution to reveal very low quantities of toxic gas concentrations. In this sense, the lock-in amplifier can be employed when the measurand shows a very small amplitude, also lower than noise level.

Commercial lock-in amplifiers, as well as ad-hoc lock-in solutions recently proposed in the literature [1, 2], typically of digital kind, have high dimensions and costs, since they are based on DSP. In this sense, especially in sensor applications, the analog kind of the signal to be revealed suggests, when the signal-to-noise ratio is about less than unity, the use of analog lock-in systems [3–6].

Traditional lock-in amplifiers (both analog and digital) have the characteristic of requiring, at power-on as well as during its working time, the manual zeroing of the output signal, related to the "in-quadrature" (90°) condition (initial phase alignment) between input and reference signals. Then, the manual activation of a switch gives a $-90°$ phase shift, achieving the required "in-phase" condition that allows to read a non-zero output voltage proportional to the mean value of the input signal, typically buried into noise.

The here presented automatic analog lock-in amplifier, which represents an advance of the topology proposed in [6] (patent pending [7, 8]), overcomes this disadvantage, being the "in-phase" condition always and continuously guaranteed by an automatic operation, ensured by a suitable feedback. Preliminary experimental measurements have been conducted on the designed system (implementing a PCB prototype based on a commercial operational amplifier as active block and precise passive components) using carbon monoxide (CO) as testing gas and FIGARO TGS2600 as commercial sensor. The achieved results have confirmed the correct functionality of the designed amplifier, as well as the system capability to reveal very small signals coming from resistive sensors with a good sensitivity and resolution improvements.

2 The Proposed Solution: The Automatic Lock-In Amplifier

The proposed analog automatic lock-in architecture, at block level, is shown in Fig. 48.1. Each block, internally, has been implemented through commercial discrete active and passive components. The system continuously provides the required "in-phase" condition by means of an automatic operation (phase self-alignment) given by suitable feedback connections. The final blocks of the proposed architecture are low-pass filters that perform a DC extraction ensuring the specific improvement of system resolution. The amplifier presents two AC inputs and three DC outputs: the two calibration outputs must be zero, in this case the measured value $V_{O,MEASURE}$ is proportional to gas concentration.

Figure 48.2 depicts an example of the measured main signals $V_{O,MEASURE}$ and $V_{O,CALIBRATION}$ (the lock-in time responses) when an input clean signal has been

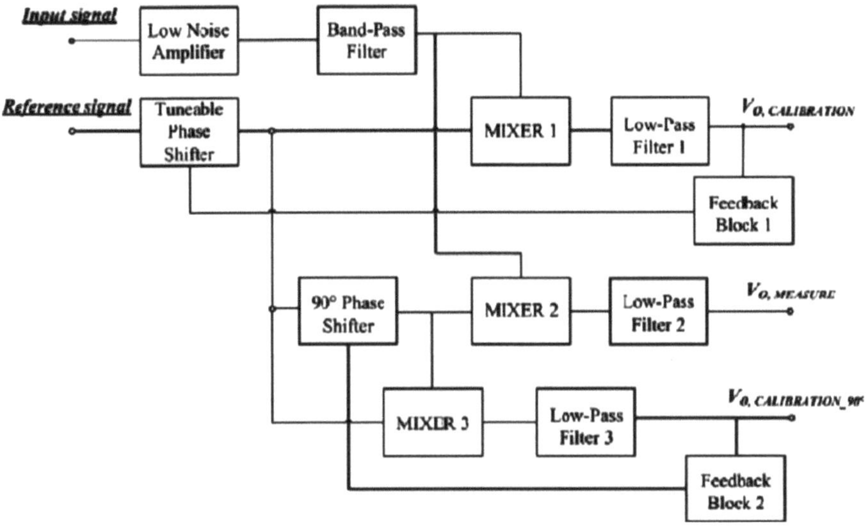

Fig. 48.1 The block scheme of the proposed automatic analog lock-in amplifier

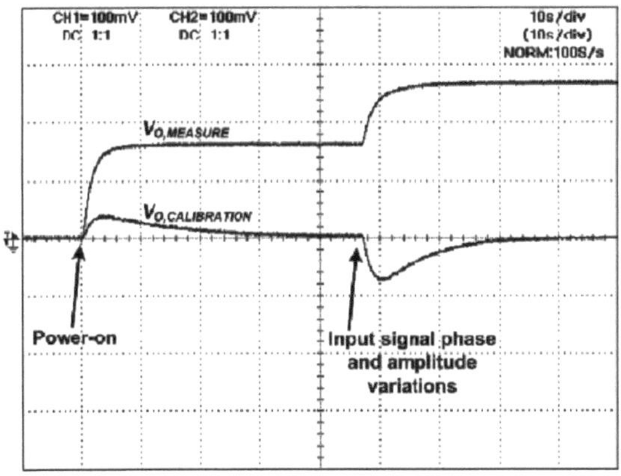

Fig. 48.2 Automatic analog lock-in time responses: system self-alignment at power-on followed by input signal phase and amplitude variations (ΔV_{IN} =2 mV; $\Delta\varphi = 20°$)

applied, showing the system self-alignment at its power-on and when amplitude (2 mV) and phase (20°) variations have simultaneously occurred.

3 Experimental Measurements

The proposed lock-in amplifier has been implemented through a PCB prototype for preliminary experimental measurements. It has been tested by the set-up scheme reported in Fig. 48.3 to detect the presence of toxic gas into a closed chamber, in

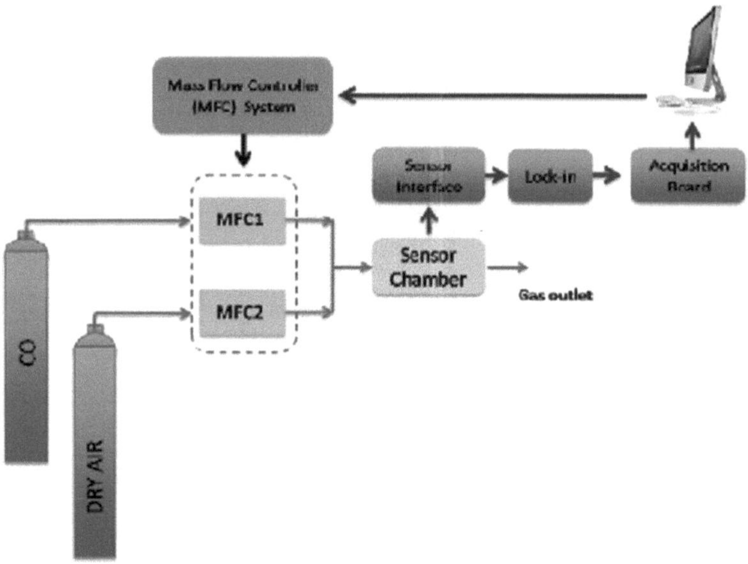

Fig. 48.3 Experimental set-up for the *CO* detection through the commercial resistive gas sensor FIGARO TGS2600

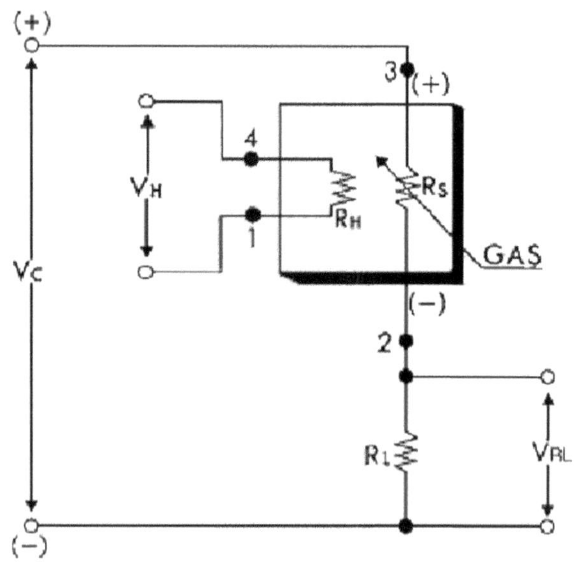

Fig. 48.4 FIGARO TGS2600 mounting scheme

particular *CO* (10, 20 and 30 ppm). The resistive gas sensor (R_S) FIGARO TGS2600 has been supplied through a 77 Hz sinusoidal voltage signal having 30 mV maximum amplitude (V_C, with a DC level of 5 V), in series with a reference load resistance, R_L, valued 10 kΩ, as sown in Fig. 48.4. The heater resistance R_H has

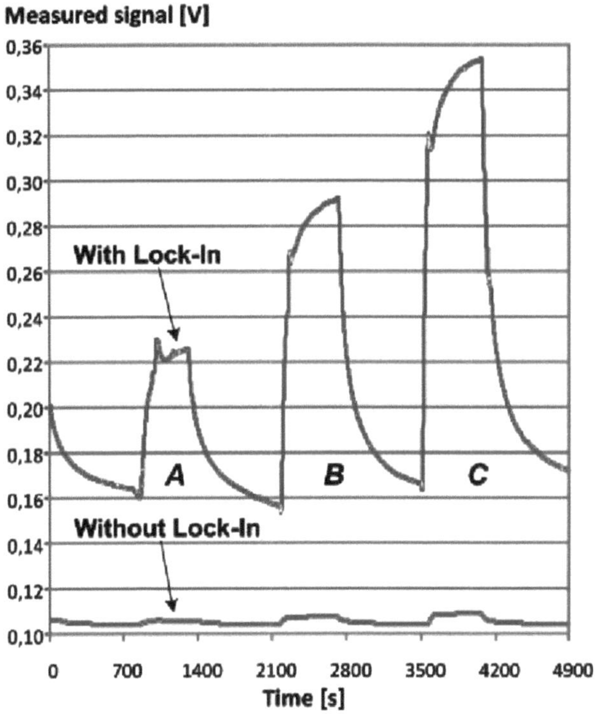

Fig. 48.5 Measured time response of the extracted DC voltage signal at the proposed lock-in output and voltage signal at the system input vs. time for different *CO* concentrations (A = 10 ppm, B = 20 ppm, C = 30 ppm)

been powered with a DC voltage level V_H equal to 5 V. In repeated measurement sessions, for 9 minutes into a closed chamber, a mixture of dry air and *CO* at different concentrations, alternated with a 14 min of dry air only, has been fluxed.

Figure 48.5 shows typical system time responses, considering both input and output DC lock-in amplifier signals for different *CO* concentrations, as detailed in Table 48.1, where the mean values of the sensor resistance have been determined over all the experimental measurements. These voltage signals have been revealed and acquired through a DAQ board, with a sampling rate equal to 1 s, allowing to estimate both the gas sensor resistance value and its variation, under the presence of different *CO* concentrations. Through a straightforward analysis of the experimental results, the sensitivity improvement given by the proposed lock-in amplifier results to be of a factor of about 80 (circuit input sensitivity ≈0.08 mV/ppm; circuit output sensitivity ≈6.5 mV/ppm), while the resolution, starting from about 5 ppm (system input resolution), has been enhanced to a theoretical value of about 0.05 ppm (system output resolution), achieving an improvement factor of about 100 for a measured noise level of about 0.30 mV.

Table 48.1 Experimental results achieved through the fabricated PCB prototype with related sensor resistance (R_S) estimation (see Fig. 48.5)

Measurement time [min]	*CO* concentration [ppm]	Mean sensor resistance $< R_S >$ [kΩ]
0–14 Initial cleaning	(Dry air only)	128
14–23 (A) Dry air + CO mixture	10	91
23–37 Cleaning	(Dry air only)	130
37–46 (B) Dry air + CO mixture	20	69
46–60 Cleaning	(Dry air only)	129
60–69 (C) Dry air + CO mixture	30	56
69–83 Final cleaning	(Dry air only)	129

4 Conclusions

The proposed fully-analog automatic lock-in amplifier has been demonstrated to be suitable for sensor interface applications and, in particular, for very low gas concentration detection. It has shown the capability to perform a phase self-alignment between the input and reference signals so to provide continuous and accurate measurements of the input signal amplitude. The designed system improves the minimal resolution of the sensor front-end, allowing the detection of very small sensor resistance variations, corresponding to very low quantity of target gas concentrations. These performances can be particularly useful in detecting some kind of toxic gases where also the presence of reduced quantities can be particularly dangerous.

Acknowledgments This work was supported by the Italian Ministry of University (MIUR) under a Program for the Development of Research of National Interest (Italian PRIN Project No 2008XZ44B8).

References

1. Marschner U, Grätz H, Jettkant B, Ruwisch D, Woldt G, Fischer WJ, Clasbrummel B (2009) Integration of a wireless lock-in measurement of hip prosthesis vibrations for loosening detection. Sensors Actuators A 156(1):145–154
2. Sonnaillon MO, Bonetto FJ (2005) A low-cost, high-performance, digital signal processor-based lock-in amplifier capable of measuring multiple frequency sweeps simultaneously. Rev Sci Instrum 76:024703(1–7)

3. Ferri G, De Laurentiis P, Di Natale C, D'Amico A (2001) A low voltage integrated CMOS lock in amplifier prototype for LAPS applications. Sensors Actuators A 92:263–272
4. Azzolini C, Magnanini A, Tonelli M, Chiorboli G, Morandi C (2008) Integrated lock-in amplifier for contact-less interface to magnetically stimulated mechanical resonators. Proc. IEEE Internat. Conference Design and Technology of Integrated Systems in Nanoscale Era, Tozeur, Tunisia, Mar 2008, pp 1–6
5. Gnudi A, Colalongo L, Baccarani G (1999) Integrated lock-in amplifier for sensor applications. Proceedings IEEE ESSCIRC, Duisburg, Germany, Sep 1999, pp 58–61
6. D'Amico A, De Marcellis A, Di Carlo C, Di Natale C, Ferri G, Martinelli E, Paolesse R, Stornelli V (2010) Low-voltage low-power integrated analog lock-in amplifier for gas sensor applications. Sensors Actuators B 144(2):400–406
7. De Marcellis A, Ferri G, Stornelli V,D'Amico A,Di Natale C, Martinelli E, Falconi C (2008) "Analog system based on a lock-in amplifier showing a continuos and automatic phase alignment", Patent No RM2008-A194, 2008
8. De Marcellis A, Di Giansante A, Ferri G, Di Natale C, Martinelli E, D'Amico A (2010) Analog automatic lock-in amplifier for very low gas concentration detection. Proceedings of Eurosensors XXIV, Linz, September 2010

Chapter 49
A CCII-Based Oscillating Circuit as Resistive/Capacitive Humidity Sensor Interface

Andrea De Marcellis, Claudia Di Carlo, Giuseppe Ferri, Carlo Cantalini, and Luca Giancaterini

In this paper, we propose a Current-Mode (*CM*) square-wave oscillator, formed by two Second Generation Current Conveyors (*CCIIs*) and some passive components, operating an impedance-to-period conversion that, instead of other solutions in the literature, is based on a current differentiation. The circuit is suitable, for example, for resistive/capacitive humidity sensor interfacing and works also for a wide oscillation frequency range (corresponding to up to six to seven variation decades of capacitive variations). It is possible to easily set its sensitivity to sensor parameters (resistance or capacitance) through external passive components. The proposed interface has been designed as an integrated solution at transistor level in a standard CMOS technology (AMS 0.35 μm) with low voltage (±1 V) and low power (430 μW) characteristics; this solution is able to properly work with integrable passive component values (resistance ≤100 kΩ and capacitance ≤100 pF), so it is also suitable for integrated portable sensor applications. In order to verify the interface validity, some experimental measurements have been performed implementing the proposed circuit through a PCB prototype utilizing AD844 as *CCII*, commercial passive sample resistors and capacitors and gas (e.g., TGS Series by Figaro) and humidity (e.g., HCH-1000 Series by Honeywell) sensors. Measurement results have shown good linearity and accuracy both for variations of floating capacitive sensors, having a baseline or changing their value in the range [pF,nF], as well as for variations of grounded resistive sensors, ranging from few kΩ to hundreds of kΩ.

A. De Marcellis (✉) • C. Di Carlo • G. Ferri
Department of Electrical and Information Engineering, University of L'Aquila, L'Aquila, Italy
e-mail: andrea.demarcellis@ing.univaq.it

C. Cantalini • L. Giancaterini
Department of Chemistry, Chemical Engineering and Materials
University of L'Aquila, L'Aquila, Italy

A. D'Amico et al. (eds.), *Sensors and Microsystems: AISEM 2011 Proceedings*, Lecture Notes in Electrical Engineering 109, DOI 10.1007/978-1-4614-0935-9_49,

1 Introduction

Oscillating circuits are applied in telecommunications, control systems, signal processing, measurement systems and, some times, in sensor interface, depending especially on their frequency operating range. These oscillators are typically implemented by using an Operational Amplifier (*OA*) as a switching current source to charge and discharge a grounded timing capacitor (i.e., a capacitive sensor), followed by a voltage hysteresis comparator [1,2]. Generally, these solutions, operating an impedance-to-period (or impedance-to-frequency) conversion, are based on a passive or active integrating cell and can be used as basic interface circuits for both resistive and capacitive sensors [2]. A limitation for these oscillators is given by the well-known finite gain-bandwidth product for the *OA*. This problem can be overcome by the use of Second Generation Current Conveyor (*CCII*) that shows good advantages in analog integrated circuit design as large bandwidth, high linearity, wide dynamic range, simple circuitry and low power consumption [3,4]. In the literature, different circuits, based on an integrating cell and developed in Current-Mode (*CM*) approach, have been proposed [5–7].

In this paper, a *CM* square-wave oscillator, based on two *CCIIs* and operating an impedance-to-period (*C-T* or *R-T*) conversion, is proposed. The circuit is suitable for resistive/capacitive humidity (or gas) sensor interfacing [8,9]. Its main operation is based on a current differentiation which allows to neglect, in the square waveform generation, the *CCII* node saturation effects that typically affect all the other solutions based on the integrating cell. Moreover, through the use of only resistive loads on the *CCII* X node, the interface does not show any limitation in a wide oscillation frequency range (up to 6–7 variation decades for capacitive variations) and it is possible to easily set its sensitivity to sensor parameters (resistance or capacitance) through external passive components.

2 The Proposed Interface: Theory and Experimental Results

The presented oscillating circuit is shown in Fig. 49.1. It is formed by six resistors, a capacitor and two positive *CCIIs*: the first, $CCII_1$, is a voltage-to-current converter, while the second, $CCII_2$, is a hysteresis current comparator, based on a *CM* Schmitt trigger. The circuit allows to neglect the Z and Y nodes saturation effects in the square waveform generation, so in capacitive sensor behaviour estimation, utilizing only suitable resistive loads on X node. In fact, the capacitive sensor is connected at Z node, so is not strongly affected by its parasitic capacitance and there are not limitations for wide variation ranges (higher than 6 decades) and high frequency (i.e., small period) values since it is possible to easily set the interface working range through several external parameters (only resistances) which allow also to set the desired sensitivity of the read-out circuit.

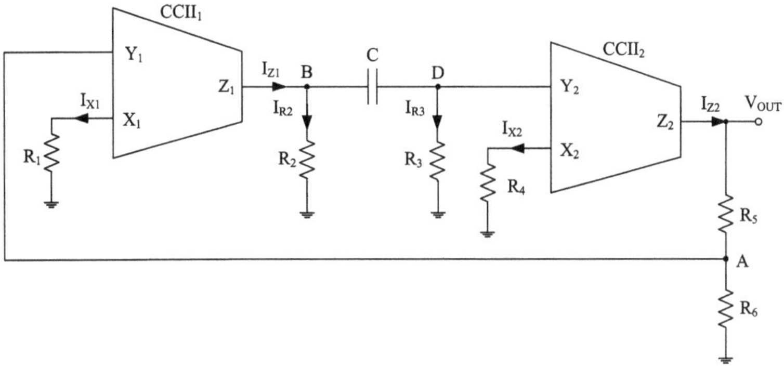

Fig. 49.1 Block scheme of the proposed resistive/capacitive sensor interface

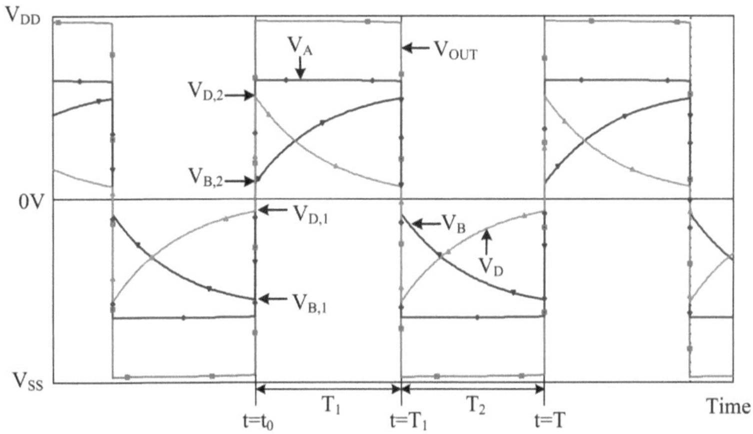

Fig. 49.2 Time responses evaluated at main interface nodes

Figure 49.2 shows the voltage signals at each node of the interface under the hypothesis of component constant values during the measuring operation. Through a straightforward analysis (see [9]), considering ideal *CCII* behavior, it is possible to achieve the following expression for the period T, revealed at V_{OUT} node:

$$T = 2C(R_2 + R_3)\ln\left[\frac{2R_2R_3R_6 - R_1R_4(R_2 + R_3)}{R_1R_4(R_2 + R_3)}\right] \tag{49.1}$$

From Eq. 49.1, for example in capacitive sensor applications, circuit sensitivity can be opportunely set by choosing suitable values of resistances R_2 and R_3, especially.

The proposed front-end topology has been designed as a complete integrated solution at transistor level in a standard CMOS technology (AMS 0.35 μm), with low voltage (±1 V) and low power (430 μW) characteristics. The proposed circuit

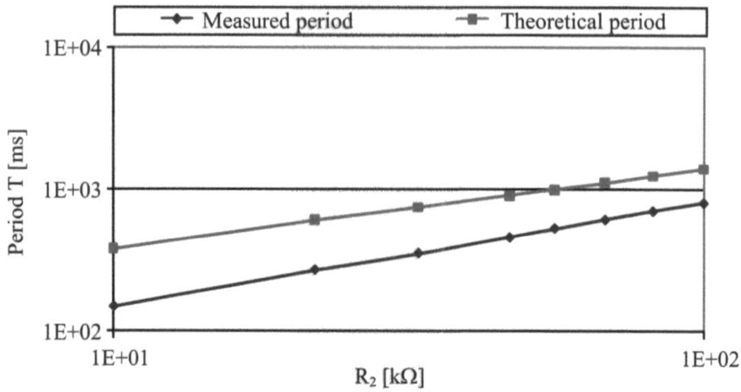

Fig. 49.3 Theoretical response (referred to ideal *CCII* behavior) and measurement results related to oscillation period of generated output square waveform vs. R_2

properly works with integrable passive component values (resistance $\leq$ 100 kΩ and capacitance $\leq$ 100 pF), so it is suitable for integrated portable applications.

Simulation results have confirmed the circuit stability for working temperature drifts (the maximum difference of the obtained oscillation period with respect to its value at the room temperature, 27°C, is lower than 3% in the whole considered range of variation, equal to [–50°C; +110°C]), showing a good linearity in a wide oscillation period range, which can be independently adjusted through either capacitive (in the range pF–μF, about six decades, for capacitors higher than 10 pF) or resistive (in the range MΩ–GΩ, about three decades, for resistors higher than 500 kΩ) external passive components. In particular, R_2 variation provides the same effects on the oscillation frequency as R_3, but for a more reduced resistive range. This constraint is due to the presence of the parasitic resistance at $CCII_1$ Z node, whose finite value limits the resistive load R_2 [9].

Then, experimental measurements have been performed implementing the circuit through a prototype *PCB* with the commercial component AD844 of Analog Devices (supplied at ±15 V) as *CCII* and using commercial passive sample resistors and capacitors, emulating both capacitive and resistive sensor behaviour. In particular, Fig. 49.3 shows the measured period variation with respect to R_2 ranging from 10 to 100 kΩ, compared to ideal value. The difference between the two curves is due to the fact the in Eq. 49.1 *CCII* non-idealities (in particular, parasitics) are not considered.

Regarding the capacitive dependence of the oscillation period, experimental results have confirmed the theoretical expectations, as reported in Fig. 49.4 (the circuit sensitivity, considering ideal *CCIIs*, is about 10 μs/pF), showing a good linearity in an oscillation period range varying *C* from 10 pF up to 10 nF. This range covers a large number of commercial capacitive sensors (e.g., pressure and humidity sensors).

Further experimental measurements have been performed employing commercial sensors, in particular capacitive humidity (HCH-1,000 Series by Honeywell)

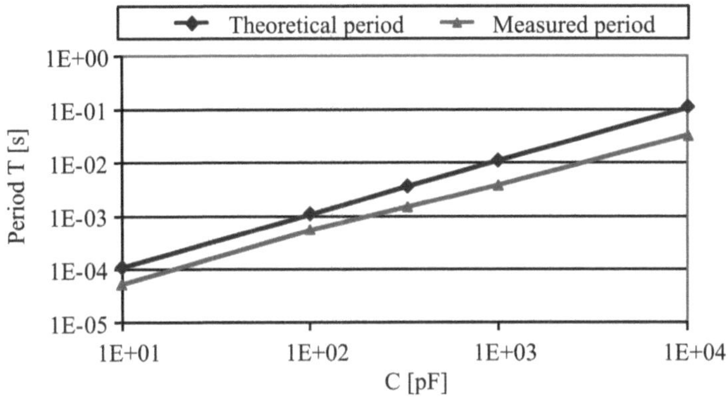

Fig. 49.4 Theoretical response (referred to ideal *CCII* behavior) and measurement results related to the oscillation period of the generated output square waveform versus. *C*

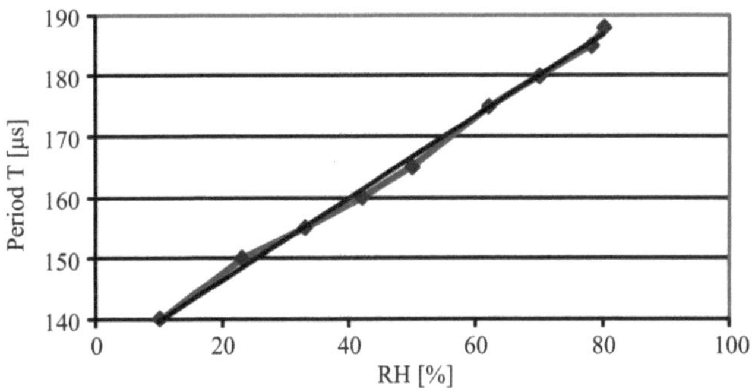

Fig. 49.5 Experimental measurements of *RH* detection through the commercial capacitive humidity sensor HCH-1000 Series by Honeywell

and resistive gas (TGS 2600 Series by Figaro) sensors. Figure 49.5 shows the period variation versus the capacitive sensor variation (i.e., *C-T* conversion), when the *RH* has been changed in the range 10–80%, properly mixing dry air with wet air in a closed and controlled chamber. In this case, the *RH* reference measurements have been achieved by a commercial thermo-hygrometer (HTD-625 High Accuracy Thermo-Hygrometer) having a resolution of about 0.1%*RH* and an accuracy of about ±2%*RH*. On the contrary, as regard the resistive dependence of the oscillation period (i.e., *R-T* conversion), the achieved experimental results have been reported in Fig. 49.6, where the resistive gas sensor provides period variations for gas concentration changes ranging from 0 up to 150 ppm. In this case, the employed gas is the *CO*, fluxed into a closed chamber with controlled concentrations. Both experimental measurements show an acceptable linearity in the oscillation period variation range.

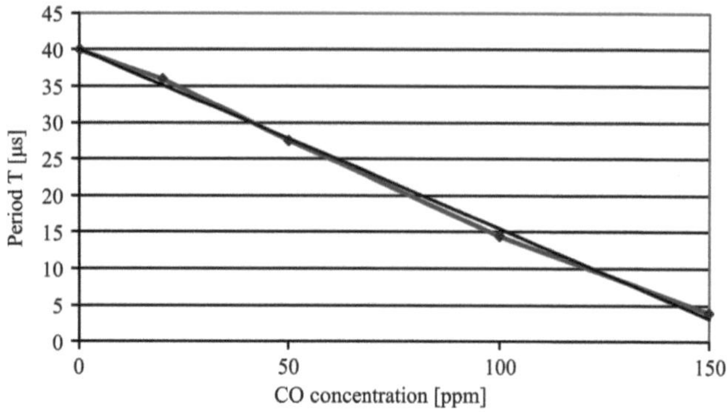

Fig. 49.6 Experimental measurements of *CO* detection through the commercial resistive gas sensor TGS 2600 Series by Figaro

3 Conclusions

In this paper, a *CCII*-based square waveform generator as first analog interface for sensor applications has been presented. The oscillating circuit is based on a differentiating cell instead of the classical integrating one, having a simple circuit topology implemented by only two positive *CCIIs*, so it is suitable for the integration on chip in a standard CMOS technology with low voltage low power characteristics. Its validity has been demonstrated through PSpice simulations for the integrated version as well as by experimental measurements using the fabricated PCB prototype and commercial gas and humidity sensors. Due to the good linearity and wide frequency range, the proposed configuration can be also considered a suitable solution for tuneable oscillators, PWL function synthesis, folding ADCs, etc..

References

1. De Marcellis A, Depari A, Ferri G, Flammini A, Marioli D, Stornelli V, Taroni A (2008) A CMOS integrable oscillator-based front end for high-dynamic-range resistive sensors. IEEE Trans Ins Meas 57(8):1596–1604
2. Haque AS, Hossain MM, Davis WA, Jr Russell HT, Carter RL (2008) Design of sinusoidal, triangular, and square wave generator using current feedback operational amplifier (CFOA). IEEE proceedings—Region 5 Technical, professional and student conference, Kansas City, Missouri, 2008, pp 1–5
3. Di Carlo C, De Marcellis A, Stornelli V, Ferri G, Tiberio D (2009) A novel LV LP CMOS internal topology of CCII + and its application in current-mode integrated circuits. IEEE proceedings—PRIME, Cork, Ireland, July 2009, pp 132–135
4. Ferri G, De Marcellis A, Di Carlo C, Stornelli V, Flammini A, Depari A, Marioli D, Sisinni E (2009) A CCII-based low-voltage low-power read-out circuit for DC-excited resistive gas sensors. IEEE Sensors J 9(12):2035–2041
5. Di Cataldo G, Palumbo G, Pennisi S (1995) A schmitt trigger by means of a CCII. Int J Circ Theor Appl 23:161–165

6. Del Re S, De Marcellis A, Ferri G, Stornelli V (2007), Low voltage integrated astable multivibrator based on a single CCII. IEEE Proceedings of PRIME, Bordeaux, July 2007, pp 177–180
7. Abuelma'atti MT, Al-Absi MA (2005) A current conveyor-based relaxation oscillator as a versatile electronic interface for capacitive and resistive sensors. Int J Electron 92:473–477
8. De Marcellis A, Di Carlo C, Ferri G, Stornelli V (2009) A novel general purpose current mode oscillating circuit for the read-out of capacitive sensors. Proceedings of IEEE IWASI, Trani, June 2009, pp 168–172
9. De Marcellis A, Di Carlo C, Ferri G, Stornelli V (2011) A CCII-based wide frequency range square waveform generator. Accepted for publication on Int J Circuit Theory Appl. doi: 10.1002/cta.781

Chapter 50
An Accurate and Simple Frequency Estimation Method for Sensor Applications

G. Campobello, G. Cannatà, N. Donato, M. Galeano, and S. Serrano

Precise frequency estimation methods for acoustic frequencies are needed in several sensor applications, however when a huge number of sensors must be monitored (e.g. sensor networks or sensor arrays) the trade-off among accuracy, speed and costs must be considered. In this paper a low-cost method for accurate frequency estimation is presented. The method can be easily implemented in a commercial microcontroller, and both analytical study and experimental results show that it is faster and more accurate than a simple FFT.

1 Introduction

Accurate frequency estimation methods for acoustic frequencies are needed in several sensor applications (implantable hearing systems [1, 2], indoor electronic nose [3], submarine eruptive activity [4]). However when a huge number of sensors must be monitored (e.g. sensor networks or sensor arrays) the trade-off among accuracy, speed and costs must be considered. In these cases the most common and cost effective solution relies on the use FFT based algorithms implemented on microcontrollers [5] but due to their limited resources (available memory, maximum clock frequency, etc.) high accuracy and high speed are difficult to achieve together.

In this paper a low-cost method for accurate frequency estimation, named NLFFE (Non-Linear Filtering method for Frequency Estimation), is presented.

G. Campobello • G. Cannatà • N. Donato • M. Galeano (✉) • S. Serrano
Dipartimento di Fisica della Materia e Ingegneria Elettronica,
DFMIE, Università degli Studi di Messina, Messina, Italy
e-mail: mgaleano@ingegneria.unime.it

A. D'Amico et al. (eds.), *Sensors and Microsystems: AISEM 2011 Proceedings*,
Lecture Notes in Electrical Engineering 109, DOI 10.1007/978-1-4614-0935-9_50,

The method can be easily implemented in a commercial microcontroller, and both analytical study and experimental results on acoustic signals show that it is faster and more accurate than a simple FFT.

2 Proposed Method

The proposed method is based on the estimation algorithm proposed by the same authors in [6] that has been simplified in order to take into account available resources of low-cost microcontroller architectures. In particular evaluation of trigonometric functions and divisions are avoided and only small size LUTs (Look-Up Tables) are used. For sake of clarity in this section we briefly review the method and some analytical results obtained in [6]. Let us indicate with s_n a generic sample of a sinusoidal signal, i.e.

$$s_n = s(nT) = A\sin(2\pi f_0 nT + \phi) \tag{50.1}$$

where T is the sampling period. By using trigonometric relations it is possible to prove that the ratio between $r_{num} = s_{n+4} - s_n$ and $r_{den} = s_{n+3} - s_{n+1}$ is

$$r = \frac{r_{num}}{r_{den}} = 2\cos\left(2\pi\frac{f_0}{f_s}\right) \tag{50.2}$$

Therefore, if the sampling frequency $f_s = 1/T$ is known, the frequency of the sinusoidal signal, f_0, can be estimated by inverting Eq. 50.2. Obviously, in order to obtain a better frequency estimation, more samples *(N)* can be used for evaluating a mean value of r before to use Eq. 50.2. More precisely given N samples we can evaluate

$$\overline{r} = \frac{1}{N-4}\sum_{i=1}^{N-4} r_i = \frac{1}{N-4}\sum_{i=1}^{N-4}\frac{s_{i+4} - s_i}{s_{i+3} - s_{i+1}} \tag{50.3}$$

It is worth nothing that, in presence of noise, the frequency estimation can be improved if ratios with small denominators, i.e. below a proper threshold B_d, are not considered for the mean.

We call sample values such that $r_{den} < B_d$ as bad values and we remove them before to evaluate $\overline{r}$. In particular, we set the threshold to $B_d = \sqrt{2}\cdot A$ so that, on the basis of the probability density function of a noisy sinusoidal signal, at least half of the values will be above the threshold and can be used to evaluate $\overline{r}$. Finally, using $\overline{r}$ we can estimate the frequency $\widehat{f}$ by inverting Eq. 50.2:

$$\hat{f} = \frac{f_s}{2\pi}\cos^{-1}\left(\frac{\overline{r}}{2}\right) \tag{50.4}$$

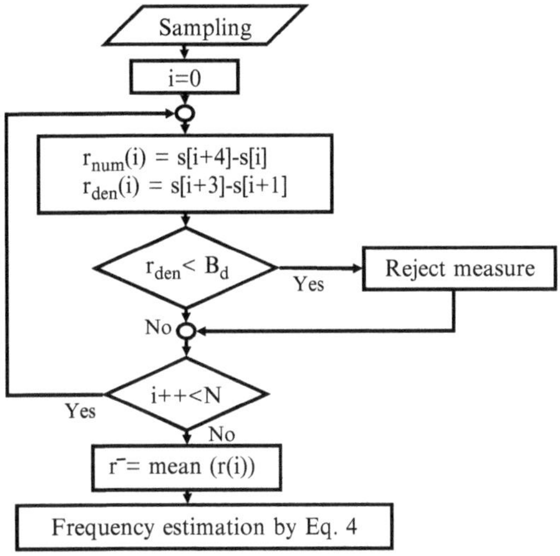

Fig. 50.1 Flow diagram of the NLFFE method

In order to avoid evaluation of the arccosine function we can restrict the frequency range so that Taylor's expansion can be used instead without a large error. For instance if the frequency range is restricted to $[f_s/4\text{-}f_s/10, f_s/4 + f_s/10]$ the maximum error due to the approximation of the arccosine function with the Taylor expansion $\cos^{-1} \approx \pi/2\text{-}x\text{-}x^3/6$ is about 0.7% (and therefore the error is less than 0.1% for the ratio $\hat{f}/f_s$). Obviously, if a lower error is needed further terms of the Taylor's expansion can be used.

In order to further simplify the implementation we avoid divisions by evaluating the ratio $1/r_{den}$ (see Eq. 50.2) with a LUT so that r can be obtained as $r = r_{num*}\, LUT(r_{den})$. Furthermore the number of samples used to evaluate $\overline{r}$ is restricted so that the factor $(N - 4)$ in Eq. 50.3 is always a power of 2.

The above algorithm is summarized in Fig. 50.1.

2.1 Performance and Complexity Analysis

Basically, without considering simplifications introduced in the previous section, the above algorithm coincides with the first step of the algorithm proposed in [6], where the authors proved that the variance of the frequency estimation is upper bounded by

$$\sigma_{nf}^2 = 0.0350 \frac{f_s^2}{N \cdot SNR} \tag{50.5}$$

where SNR is the signal-to-noise ratio of the sampled signals, and the maximum error in frequency estimation can be estimated as

$$\Delta f_{ME} = 3\sigma_{nf} = 0.561 \frac{f_s}{\sqrt{N \cdot SNR}} \tag{50.6}$$

By comparing Δf_{ME} of the proposed method with the maximum error of an N-point FFT (i.e. $\Delta f = f_s/(2\ N)$, see Ref. [7]) it is straightforward to prove that the error of the proposed method is smaller if $SNR > 1.25\ N$.

Regarding complexity of the proposed method, let us observe that in the worst case (i.e. if there are not bad values) a maximum of $3(N - 4)$ additions and $(N - 4)$ multiplications are needed to obtain $\bar{r}$; further three multiplications and two additions are needed to evaluate $\cos^{-1}(\bar{r})$ with the Taylor expansion; and only one multiplication is needed to finally estimate $\hat{f}$ as in Eq. 50.4 (considering that the constant $f_s/(2\pi)$ is stored within the microcontroller's memory).

Therefore the proposed algorithm needs about N real multiplications and about $3\ N$ real additions.

When compared with a N-point FFT the number of multiplications needed is reduced by a factor $2log_2(N)$. In fact a N-point FFT needs about $(N/2)log_2(N)$ complex multiplications (see [7]) that in a microcontroller must be implemented as $2Nlog_2(N)$ real multiplications.

So, if we consider for instance $N = 128$ samples, the proposed method is 14 time faster than a FFT and achieve a greater accuracy if the SNR is greater than 22 dB.

3 Experimental Results

The proposed method has been implemented in an ATMega8 microcontroller which integrates an 10-bit ADC. However, to simplify memory management and improve code size and speed, only the most significant 8 bits are used. As case study two sampling frequencies have been used: 4430.77 Hz (for sinusoidal signals with a frequency between 600 Hz and 1,400 Hz) and 17,723 Hz (for sinusoidal signals with a frequency between 2,600 Hz and 5,840 Hz). Frequencies have been estimated using the proposed method considering $N = 128$ samples and obtained results have been compared with the results given by a 128-point FFT and a 65,536-point FFT, both implemented off-line by importing samples on a Personal Computer. In particular, frequencies obtained by the 65,536-point FFT are used as reference values (i.e. actual frequencies) to evaluate the absolute errors.

Experimental results are reported in Fig. 50.2 where the absolute errors of the proposed method and the 128-point FFT are shown. As it is possible to observe the proposed method has a greater accuracy. More precisely, for the first sampling

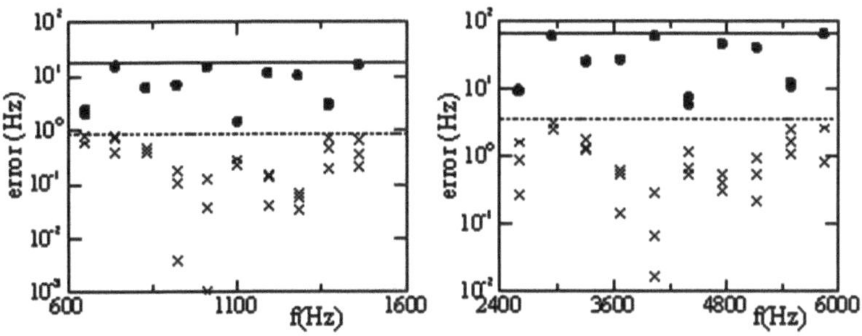

Fig. 50.2 Experimental results of the proposed method (NLFFE) compared with a 128-point FFT. *Dots* are FFT errors, *cross* are NLFFE errors. *Continuous lines* represents theoretical FFT maximum errors *dashed lines* represents theoretical NLFFE maximum errors

frequency (4430.77 Hz) the maximum error on 30 measures (ten frequencies have been measured, each for three times) is less than 1 Hz (in comparison to 17.31 Hz of a 128-point FFT), and for the second sampling frequency (17,723 Hz) the maximum error is 3.5 Hz (in comparison to 69.23 Hz). Let us observe that the maximum error can be predicted by Eq. 50.6. In fact the SNR can be estimated as $6m \approx 48dB$ (i.e. considering the 6 dB-law of quantized signals) and on the basis of Eq. 50.6 we have 0.87 Hz as maximum error when $f_s = 4430.77$ Hz and 3.5 Hz as maximum error when $f_s = 17{,}723$ Hz.

4 Conclusion

Further experimental activities are in progress. In particular we are using the proposed method in some experimental setups where acoustic and quartz-based sensors are involved.

References

1. Ko WH et al (2009) Studies of MEMS acoustic sensors as implantable microphones for totally implantable hearing-aid systems. IEEE Trans Biomed Circuits Syst 3:277
2. Lee J et al (2009) A surface micromachined MEMS acoustic sensor with X-shape bottom electrode anchor. In: IEEE Sensors, Christchurch, 2009
3. Yao D (2009) A gas sensing system for indoor air quality control and polluted environmental monitoring. In: IEEE NANO Organizers, Genoa, 2009
4. Matsumoto H et al (2010) Hydroacoustics of a submarine eruption in the Northeast Lau Basin using an acoustic glider. In: IEEE Oceans, Sydney, 2010

5. Ong KS-H, Yue S-P, Ling K-V (2010) Implementation of fast fourier transform on body sensor networks. In: ICBSN (International Conference on Body Sensor Networks), Washington, DC, 2010
6. Campobello G, Cannatà G, Donato N, Famulari A, Serrano S (2010) A novel low-complex and low-memory method for accurate single-tone frequency estimation. In: 4th IEEE international symposium on communications, control and signal processing (ISCCSP 10), Limassol, Cypres 2010
7. Ifeachor E, Jervis B (2002) Digital signal processing: a practical approach. Prentice-Hall, Harlow/New York

Chapter 51
Compact Low Noise Interfaces for Multichannel MEMS Thermal Sensors

P. Bruschi, F. Butti, and M. Piotto

In this work a novel architecture for the design of compact instrumentation amplifier is described. The low offset and low noise characteristics of the proposed amplifier make it particularly suitable for interfacing thermopile-based MEMS sensors. The circuit consists in a fully differential 2nd order low pass Gm-C filter, properly modified to provide gain and incorporate chopper modulation. The validity of the approach is proven by means of simulations performed on a prototype designed with a commercial CMOS process.

1 Introduction

Thermal sensors represent a successful example of Micro Electro-Mechanical Devices (MEMS). These sensors convert a physical or chemical quantity into a temperature difference that develops across distinct points of the same silicon chip. The temperature differences are easily measured with virtually no offset using thermocouples [1]. In order to fully exploit the advantages of MEMS thermal sensors it is necessary to read the thermopile output voltages with a resolution of the order of a few microvolts. Instrumentation amplifiers (in-amps) are the most versatile blocks for interfacing thermal sensors, due to their differential input, high input resistance and low noise/low offset. The continuous trend towards fully integrated systems, including several sensors [2] and the related readout electronics

P. Bruschi (✉) • F. Butti
Dipartimento di Ingegneria dell'informazione: Elettronica, Informatica, Telecomunicazioni, Università di Pisa, Pisa, Tuscany, Italy
e-mail: p.bruschi@iet.unipi.it

M. Piotto
CNR IEIIT – Pisa, Pisa, Tuscany, Italy

A. D'Amico et al. (eds.), *Sensors and Microsystems: AISEM 2011 Proceedings*, Lecture Notes in Electrical Engineering 109, DOI 10.1007/978-1-4614-0935-9_51,

on the same chip, urges the development of compact instrumentation amplifiers for parallel multi-sensor interfacing.

In this communication, we propose an instrumentation amplifier based on the chopper modulation technique for reaching the required characteristics of small input offset and noise voltage. Differently from the classical chopper amplifier architecture, that relies on an output low pass filter for rejecting the chopped offset (chopper ripple), we propose to embed chopper modulation into a 2nd order Gm-C filter, properly modified to provide gain. In this way, the intrinsic filtering properties of the amplifier are used to (1) cancel the chopper ripple and (2) limit the bandwidth of the readout channel, allowing direct connection of the amplifier output to a low sampling rate ADC. Thanks to the chopper modulation, the filter can be designed with relaxed noise and offset constraints with huge benefits in terms of area occupation with respect to classical schemes [3].

2 Amplifier Description

The block diagram of the filter is shown in Fig. 51.1. It is composed by two Gm-C integrators, whose unity gain frequencies are ω_{01} and ω_{02}, and a resistive attenuator of gain $\beta < 1$.

The overall gain of the amplifier is equal to $A_0 = \beta^{-1}$, while the cut-off frequency and quality factor of the stage are given by $f_c = (\beta\omega_{01}\omega_{02})^{1/2}$ and $Q = (\omega_{01}\beta/\omega_{02})^{1/2}$, respectively. The in-amp has been designed according to a fully-differential architecture, in order to obtain a differential input and facilitate the implementation of the modulators. The structure of the differential integrators is shown in Fig. 51.1 (right), where a transconductor provided of two differential ports has been used.

The amplifier has been implemented using CMOS devices. The fully differential transconductors used in Int1 and Int2 are based on input pseudo differential p-MOSFET pairs, operating in saturation region. It can be easily demonstrated that most of the noise and offset contribution in the filter pass-band comes from Int1. Therefore, Int1 topology has been devised to meet the strict specifications dictated by the extremely low signal levels of thermoelectric sensors. In order to reject the offset voltage and low frequency noise, chopper modulation has been

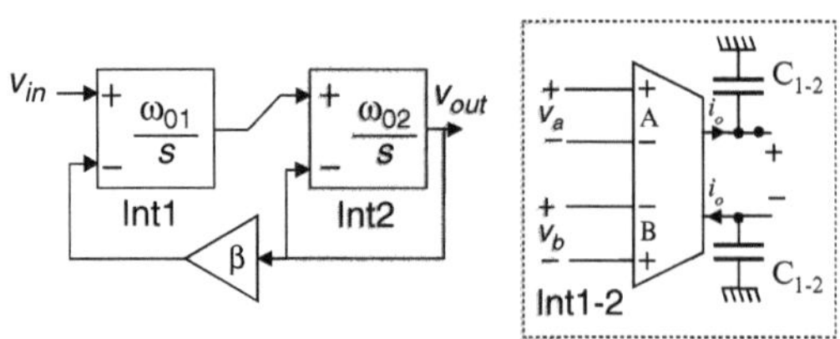

Fig. 51.1 Block diagram of an integrator based second order low pass filter providing the gain $A_0 = 1/\beta$ (*left*). Fully differential implementation of the integrators in Gm-C architecture (*right*). Capacitors used in Int1 and Int2 are indicated with C_1 and C_2, respectively

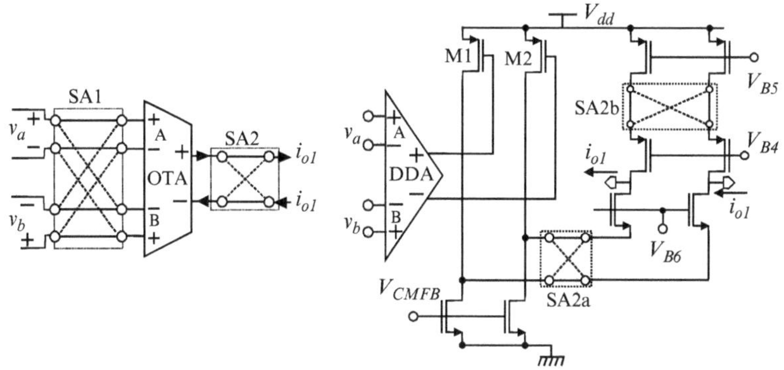

Fig. 51.2 (*Left*) Block diagram of the scheme used to apply chopper modulation to the INT1 integrator and (*right*) schematic view of the block OTA including also the output modulator SA2

applied to the transconductor used in this block. Figure 51.2 (left) shows how this is accomplished by means of proper switch arrays SA1 and SA2. In particular, SA1 also operates swapping of the input ports in order to reduce gain inaccuracy due to device mismatch with a principle similar to that proposed in [4]. Figure 51.2 (right) shows a schematic view of the block OTA, including also the output modulator SA2. The DDA (difference differential amplifier) is introduced as a preamplifier in order to relax the thermal noise constraints of the transconductor, which, are not affected by chopper modulation. A preliminary prototype based on this architecture has been described in [5]. In this work, we propose a version with similar noise characteristics but with lower power consumption. The improvement has been obtained by modifying the DDA with the use of a telescopic cascode topology and input devices biased in weak inversion. This approach was particularly effective since the DDA is the block that uses the largest fraction of the input current, due to the strict thermal noise specifications.

3 Prototype Design

The proposed prototype has been designed using the 0.32 μm–3.3 V CMOS subset of the STMicroelectronics BCD6s process. The dc gain was set to 200 with a cut-off frequency programmable over the 100–500 Hz interval in four steps by digitally varying capacitors C_1 and C_2. All the simulations in rest of the paper refer to a cut off frequency of 200 Hz and a quality factor $Q = 0.7$, obtained with $C_1 = 32$ pF and $C_2 = 16$ pF. The total supply current was 300 μA for a V_{dd} of 3.3 V. The amplifier was carefully dimensioned to obtain an input noise voltage power spectral density (PSD) lower than 30 nV/sqrt(Hz). The result is shown in Fig. 51.3 (left) were the total output PSDs is plotted together with the individual contribution of Int1 and Int2. The corresponding input referred PSD is 25 nV/sqrt(Hz).

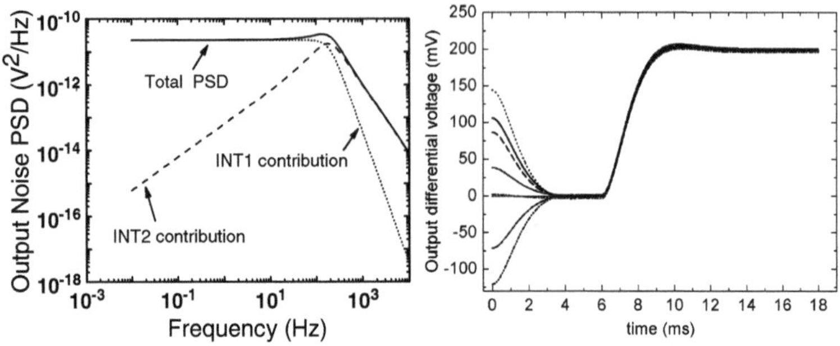

Fig. 51.3 Calculated output noise voltage spectral density and individual contributions of INT1 and INT2 (*left*). Monte Carlo transient simulations of the amplifier response to a 1 mV step (*right*)

Figure 51.3 (right) shows the simulated transient response of the amplifier to a 1 mV input step. The figure shows several Monte Carlo runs, which start from different initial voltage due to the large static offset. The latter is recovered by the effect of the chopper modulation with a maximum residual output offset of 1 mV, corresponding to an input offset of 5 μV. The gain error is less than 0.2% for all curves. Such an excellent result derived by the adoption of the port swapping approach described above. Application of standard chopper modulation technique, consisting in using distinct modulators for the signal and feedback port of the DDA, produced a gain error an order of magnitude larger.

A series of transient simulations has been performed to verify the correctness of the noise prediction. To this aim several NOISETRAN simulations, operated by activating the noise sources of all devices, have been run and the output noise *rms* voltage has been calculated. The result was in agreement with the output PSD of Fig. 51.3 (left).

The cell area, estimated by summing up the capacitor areas and the MOSFET areas, is slightly smaller than 0.16 mm^2. Due to the large dimensions of the circuit devices, the additional area required in the layout design phase for the interconnections and device spacing can be considered negligible.

4 Conclusions

The results of electrical simulations performed on the designed prototype confirmed that, in terms of input noise PSD and cell area, the proposed amplifier is practically equivalent to the previous version. The whole cell, including also the clock generator can be contained into a 400 × 400 μm^2, including the filter capacitors. On the other hand, thanks to the topology and biasing optimization, the supply current of the proposed version could be reduced to one fourth of the original amplifier The relatively low power consumption and the compactness of the cell allow several independent readout channels to be placed on relatively small chips, to accomplish parallel reading of sensor arrays.

Acknowledgments The authors would like to thank STMicroelectronics for providing the BCD6s process design kit.

References

1. Baltes H, Paul O, Brand O (1998) Micromachined thermally based CMOS microsensors. Proc IEEE 86:1660–1678
2. Alandry B,Dumas N, Latorre L, Mailly F,Nouet P (2009) A MEMS-based multi-sensor system for attitude determination. In: Proceedings of DTIP 09, Rome pp 261–264, Apr 2009
3. Ng KA, Chan PK (2005) A CMOS analog front-end IC for portable EEG/ECG monitoring applications. IEEE Trans Circuits Syst 52:2335–2347
4. Falconi C, Faccio M, D'Amico A, Di Natale C (2003) High-accuracy instrumentation amplifier for low voltage low power CMOS smart sensors. In: Proceedings of ISCAS'03, Bangkok, pp III534–III537, May 2003
5. Butti F, Dei M, Piotto M, Bruschi P (2010) A compact instrumentation amplifier for MEMS sensor interfacing. In: Proceedings of AVLSIWS, 2010, Pavia, pp 143–148, Sep 2010

MIX
Papier aus verantwortungsvollen Quellen
Paper from responsible sources
FSC® C105338

If you have any concerns about our products,
you can contact us on
ProductSafety@springernature.com

In case Publisher is established outside the EU,
the EU authorized representative is:
Springer Nature Customer Service Center GmbH
Europaplatz 3, 69115 Heidelberg, Germany

Printed by Libri Plureos GmbH
in Hamburg, Germany